DESIGN AND THERMAL PERFORMANCE

DESIGN AND THERMAL PERFORMANCE

BELOW-GROUND DWELLINGS IN CHINA

Gideon S. Golany

DELAWARE

Newark: University of Delaware Press
London and Toronto: Associated University Presses

Associated University Presses
440 Forsgate Drive
Cranbury, NJ 08512

Associated University Presses
25 Sicilian Avenue
London WC1A 2QH, England

Associated University Presses
P.O. Box 488, Port Credit
Mississauga, Ontario
Canada L5G 4M2

Library of Congress Cataloging-in-Publication Data

Golany, Gideon.
    Design and thermal performance.

    Bibliography: p.
    Includes index.
    1. Earth sheltered houses—China—Design and construction.   2. Earth sheltered houses—China—Thermal properties.   I. Title.
TH4819.E27G65   1990   690'.8   88-40425
ISBN 0-87413-362-9 (alk. paper)

The paper used in this publication meets the requirements
of the American National Standard for Permanence of Paper
for Printed Library Materials Z39.48-1984.

PRINTED IN THE UNITED STATES OF AMERICA

To Raniero Corbelletti, who will be remembered as a dedicated educator, a congenial colleague, and a humane, friendly, and tolerant gentleman. Raniero Corbelletti was professor and head of the Department of Architecture. It was he who introduced me to Penn State University and supported my scholarly endeavors.

CONTENTS

FIGURES

TABLES

PREFACE

Over the past decade, there has been increased interest throughout the world in below-ground, or earth-sheltered space. Historically, there have been a variety of uses for this type of space on an international basis, and some countries have even specialized in one or another aspect of its use. In Japan, for example, shopping centers have recently been developed below ground, one of which accommodates more than 800,000 users per day. In the United States, there has been increasing use of earth-sheltered space for personal habitats, as well as for below-ground schools, libraries, refrigeration plants, and other industrial facilities. Canada, Sweden, Norway, and other developed countries have adopted similar projects. In addition, knowledge of the ancient experience of below-ground space utilization will become increasingly important in relation to the study of modern-day land uses.

My study of the global achievements of the ancients focused on three significant regions in the world where below-ground space communities have been in continuous existence for thousands of years:

1. Southern Tunisia, bordering northern Sahara, where there are more than twenty below-ground communities. I conducted thorough field research in this region and the manuscript of my findings has been published by the University of Delaware Press.
2. The Cappadocia area, located in central Turkey, some four hundred kilometers southeast of the capital city of Ankara. In this region, where the climate is dry and warm in the summer and cool with very low precipitation in the winter, there exist more than forty below-ground communities. These settlements have been in continuous use since 4000 B.C.—and I am currently investigating.
3. The loess soil plateau of northern and north-

western China. In the five provinces of Shaanxi, Shanxi, Gansu, Henan, and Ningxia Hui Autonomous Region, it is estimated that between thirty and forty million people are living today in below-ground dwellings. This book concerns their region.

There is a great need to thoroughly understand the wealth of historical experience of below-ground space usage within the context of its design configuration and its relation to soil behavior and heat gain and loss in order to improve modern design and development. It is essential to adapt this ancient "laboraboratory knowledge tpresent-day needs and design norms. The prime goals of this study are to improve existing design and achieve better earth-sheltered habitats throughout the world.

The research for this manuscript was supported financially and otherwise by the National Academy of Sciences, Washington, D.C., through the Committee for Scholarly Communication with the People's Republic of China. My activities on this subject started in 1979, and later I received the award from the National Academy of Sciences. The field surveys began in the summer of 1984. During my field research, I was associated with Xian Institute of Metallurgy and Construction Engineering, Xian, Shaanxi province; Tongji University, Shanghai; and the Chinese Ministry of Metallurgical Industry, Beijing, who sponsored the project. I have spent a total of one year in the People's Republic of China over the course of four trips.

The focus of my field research involved the following: selecting and researching twelve cave dwellings in different geographical regions of four provinces of China as a pilot study on Chinese cave dwellings; interviewing a large number of users of below-ground space, and also professionals and others acquainted with the subject in China; referring to the limited amount of literature on this subject in the Chinese language, and the even more limited amount in English, and having a substantial number of articles and books translated from Chinese to English.

The research treatment of each of the twelve pilot dwellings was organized according to the following system:

1. Taking temperature measurements using dry- and wet-bulb thermometers, every hour for a period of twenty-four consecutive hours, both in summer and in winter, at five to six different sites in each pilot dwelling. The standard procedure was to select for measurement three or four below-ground rooms of different orientation, character and design, plus the patio that is a customary part of these structures, as well as an open space outside. The resulting temperature and relative humidity readings were plotted graphically by computer to display the comparative summer and winter results. This information was analyzed and conclusions drawn.

2. Interviewing the occupants thoroughly (through an interpreter) about their perceptions of their houses during different seasons with regard to thermal performance, comfort, basic attitudes and feelings, and such matters as the construction process, price of the structure, labor involved, soil behavior, practicality, and other factors.

3. Full mapping of each house, including plan preparation, cross sections, perspectives, views, and the like. The field draftwork was then prepared in final form for publication.

4. Taking comprehensive and extensive photographs of each house from which bird's-eye-view drawings were reconstructed to support the analyses.

In addition to the twelve cave dwellings, a survey and data collection were undertaken in other areas that accommodate below-ground dwellings. There are such regions with 80 to 90 percent of the population living below ground.

The Chinese concept of living below ground is complex and results from a variety of causes:

1. Climate. Residences are cooler in summer and warmer in winter, thus representing an energy savings.

2. Conservation. A dwelling saves land for agricultural purposes.

3. Economy. It is quicker to build and cheaper by 50 percent than building above ground;

4. Simplicity of construction. A below-ground house requires almost no building materials that are not available on or near the site itself, and the farmer-builder can use digging implements he already has for agricultural purposes.

5. Security/protection and maximum privacy.

To the best of my knowledge, this is the first comprehensive research of its kind to be prepared on Chinese cave dwellings and published in the English language. In any case, the Chinese have only recently become aware themselves of the significance of the topic, and the Architectural Society of China has appointed a national Investigation Group to study it.

Like any other field research project, this one also had its limitations and strengths. The major limitations were diverse, and included technical questions as well as cultural concepts. Only recently have a limited number of articles on the subject of Chinese cave dwellings been published in the Chinese language, and most of these are general in nature rather than research products, although there were a few published Chinese research works which were of reasonable quality. I have had almost all of these articles translated to English and have made reference to them in this book. Also, there was no published statistical data on Chinese cave-dwelling populations and their housing. Most importantly, there were no materials on the thermal performance of such dwellings. There was a need to collect such data independently. In some case, I received data from some local officials that I combined with my data to create an overall picture.

Another limitation was the large scale of the region that I investigated and the limited availability of transportation in some areas. There was a need to visit some areas that were out of bounds for foreigners at the time of my research and special permission was obtained from the authorities who were very helpful. Privacy in some of the dwellings where I measured temperatures during the day and night in different rooms was of concern and had to be worked out with the residents. Although language and cultural differences usually constitute a major communication problem, my difficulties were overcome by extensive homework and with the help of my Chinese assistant. The extensive amount of temperature data was limited to one diurnal summer and winter season. I believe that the cases, which were selected carefully, and the days of temperature measurement do represent the basic conditions of the cave dwelling's thermal performance. ·It would have been impossible both financially and logistically to collect such data for all the year or for some longer number of selected days. Because of the large amount of temperature performance information that was collected, I have presented only a limited number in the graphs. I preferred to introduce every studied case singularly to achieve effective communication with the reader.

Regardless of the above-mentioned limitations, there were many encouraging conditions for the field research and its preparation for publication. First, and most important, was the grant awarded to me by the National Academy of Sciences to cover expenses for this twelve-month field research in China. A key to facilitating the field work was the approval of the Chinese authorities for my field survey and all the logistic arrangements necessary to it, without which the field survey would have been impossible. The arrangement for a Chinese assistant to escort me throughout all the field survey became vital for recorded interviews, data collection, temperature measurement, and for logistic needs. At each site where I arrived, the local authorities were extremely helpful to my field survey. My many meetings with the newly appointed Chinese National Investigation Group of Cave Dwellings, headed by the prominent architect, Ren Zhenying, and its subcommittees of the four provinces were instrumental to me for the content of our discussions and for arranging the logistics of the field survey. Last, but not least, was the extensive homework on the Chinese cave dwelling case and China itself, which I conducted over a period of many years prior to my first visit to China in 1984, and my similar field survey experience on below-ground dwellings conducted in 1982 in southern Tunisia bordering the northern Sahara.

Many individuals and institutions, both in China and the United States, were involved with conducting this research and preparing it for publication. They all deserve my appreciation and gratitude. My thanks are due to Qiao Zhen, student of the Department of Architecture, Xian Institute of Metallurgy and Construction Engineering, who was my faithful assistant during the field survey of the twelve dwellings. My appreciation also to the students in my Department of Architecture at Pennsylvania State University who helped in the final drafting of the large number of drawings included in this book, graduate students Wu Hua and Deng Dong and senior undergraduate, Claudia Albertin; to Loukas Kalisperis, who prepared all the graphs through our ACL-COMCAD computer laboratory and to Professor Pier Bandini, Director of ACL-COMCAD, who provided the facilities; to Luke Leung, student in Architectural Engineering, and to Dr. Y. S. Lee of Tongji University, Shanghai, who translated much material from Chinese to English. My thanks also to Wang Jianzhong, art student at Henan University at the time, who was dedicated to gathering valuable information for my research and to executing some of the drawings.

The preparation of this book for publication was supported financially by Dr. Charles Hosler, vice president for research and dean of the Graduate School of Pennsylvania State University. It was his decisive support that brought this and my other China research manuscripts to realization and I am grateful for his full encouragement. I am also thankful to the Laboratory for Environmental Design and Planning of the College of Arts and Architecture, which was also helpful at a critical stage of this publication. Professor Raniero Corbelletti, former acting dean of the College of Arts and Architecture, who reviewed all the drawings and provided me with valuable comments; Dr. Robert Holmes, former dean of the college; and Professor Richard Grube, former acting head of the Department of Architecture; were all most encouraging during the later part of the project, and they deserve my deep gratitude. My appreciation is extended to my editorial assistant, Grace Perez, for her perseverance and attention to detail in working on this volume from its early stages, and to Linda Gummo who dedicated her time and effort to the careful typing of this manuscript.

I am especially grateful to Dr. James Moeser, Dean of the College of Arts and Architecture, for his continuous encouragement and for creating the necessary environment to support my scholarly work.

To all, my grateful thanks. Without them this volume would not have been possible.

ACKNOWLEDGMENTS

I would like to thank those who helped me in the realization of this research, both before and during my stay in China:

Robert Geyer, director of the National Program for Advanced Study and Research in China of the Committee on Scholarly Communication with the People's Republic of China at the National Academy of Sciences, Washington, D.C., and his program assistant, Jeff Filcik.

The staff of the Chinese Ministry of Metallurgical Industry, Beijing, who sponsored my research project during my stay in China. A large number of their key personnel facilitated every detail of my research plans. Without their perceptive help this research could not have been realized. Among them: Wang Zu-Cheng, director of the Education Department; Li Wenjian, vice director of the Education Department; Yang Song-Tao, engineer, Foreign Affairs Department; Li Fu-Qin, translator, Division of Higher Education.

Most helpful were the many staff members of the American Embassy in Beijing: Dr. Jack L. Gosnell, counselor for Science and Technology and his assistants, Christopher J. Marut and Douglas B. McNeal; the first secretary and cultural affairs officer, Dr. Karl F. Olsson; and Lynn H. Noah, Counselor for Press and Cultural Affairs.

The Architectural Society of China, which, along with the Ministry of Metallurgical Industry, was effective in managing my itinerary. Among the many helpful members: architect Ren Zhenying, head of the Investigation Group on the Chinese Cave Dwellings; Huang Sin-Fan, director of the Academic Affairs Division; Gong Deshun, former Secretary of the ASC; Xi Jingda, director of the Department of Foreign Affairs; and Jin Oubu and Zeng Jian of the ASC; and Chen Zhanxiang of the China Academy of Urban Planning and Design.

Also the many persons from Xian Institute of Metallurgy and Construction Engineering: president and Professor Zhao Hongzuo; Zhang Guang and Zhang Peixul of the Department of Foreign Affairs; members of the Department of Architecture, Professor Guang Shi-Kui, head, and Professors Zhang Si-Zan, Xia Yun and, last but not least, my friend, Professor Hou Ji Yao, who was most helpful with advice on arranging my research plan.

The largest number of persons at Tongji University in Shanghai, among them: Professor Hou Xue Yuan, head of the Department of Geotechnical Engineering, and Dr. Y. S. Lee of the Underground Structure Group; and my assistants during additional field research in Shanghai, Su Yu and Peng Fristo.

Others at Beijing University of Iron and Steel Technology: president and Professor Wang Run; Professor Yu Zong-sen, dean of the Faculty of Physical Chemistry; Professor Huang Wu-Di, provost and director of our newly initiated Interdisciplinary Graduate Program on Environmental Design; Professor Qiao Duan; and Dali Yang and Zhao Yong-Lu, interpreters.

And in Shanxi Province: architect Zuo Guo Bao of the Architectural Scientific Academy of Shanxi; and Lin Yi Shan, director of the Institute of Building Science; Li Minhong and Qiao Shuong Wong, both of the Taiyuan City Construction Committee; Jia Kuen Nan of the Architectural Society of Linfen Region; and Lian Peng, interpreter with the Foreign Affairs Office of Linfen Region.

In Shaanxi Province: Wang Zheng Ji, head of Fen Hue Team, Liquan County, Wang Yang Chang, Leader in the Foreign Affairs Division, and Li Jian Qien, architect; from Yan'an City, Wei Tan King, construction engineer of Yan'an Region, and Zhang Zong Qiang of the Foreign Affairs Office.

In Gansu Province: Ren Zhenying, chief architect of Lanzhou City, Nan Ying Jing, translator and engineer of Lanzhou City; Jing of the Qingyang Regional Government; Wang Jiu Ru of the Architectural Society of Qingyang Region and head of the Office of Urban and Rural Construction and Environmental Protection.

In Henan Province: Professor Shirley Wood, dean of the Foreign Language Department and honorary director of Graduate Studies, Henan University, Kaifeng, was most helpful, as was artist Wang Jianzhong of Henan University, who assisted me greatly before and during my stay in China; in Zhengzhou City, architect Lee Lian Shing, Architectural Society of Henan, and Chow Pei Nan of the Architectural Academy of Zhengzhou City; in Gong Xian County, Zhang Chueng Rueng and Lee Jin Lu, engineers with the Department of Construction; and in Luoyang City, Lee Chuan Zhe and Wang An Min of the Architectural Society of Luoyang Region.

And finally at Tsinghua University, Beijing: professor Li Daozeng, dean of the Department of Architecture, and Professors Cai Junfu and Wu Huanjia, both of the Department of Architecture.

To all, my thanks and deep appreciation for their perceptive help and warm hospitality.

DESIGN AND THERMAL PERFORMANCE

1

DESIGN AND THERMAL PERFORMANCE

Basics of Chinese House Design

To understand cave dwelling design we must observe the principles of Chinese house design. The traditional perception of the Chinese house has evolved over time into a symbolic national dwelling pattern. Many forces have contributed to the formation of this pattern, among them environment (sand, wind, desirable view, and the like), social (privacy, intimacy, confinement, and preservation of a strong family institution), economics (use of local materials, such as brick or adobe), and practical considerations (the average Chinese is a very pragmatic individual, yet he has a highly developed sense of art and aesthetic correctness).

In most cases, the Chinese house is the synthesis of practicality and aesthetics, and as such is highly significant. Its basic principles are composed of the following elements:

1. Confinement. The house itself is structurally defined to make it distinctive and isolated from its surroundings. This isolation is achieved by erecting walls or by using the back of a building to confine the enclosure, or both. The wall usually exceeds the height of the average person to eliminate viewing into or out of the premises.
2. Form. This is either square or rectangular and usually established by the placement of the various buildings to form an overall configuration. A theoretical line across the patio from south to north is enforced by the total layout of the buildings.
3. Courtyards. These are centrally located in the complexes, usually not covered, and are sur-

rounded by the buildings or walls. All the buildings of a house face a courtyard. Sometimes a pavilion or other lightweight structure serves as a transition from one courtyard to another. The patios thus provide privacy and an intimate meeting area for family and guests, and they are the center of family social activities. The main entrance usually leads to a courtyard and thus only indirectly to the house itself, with a wall or a building in the foreground preventing visitors from looking directly into the house from the entrance area.
4. Orientation. Consideration has always been given, even by urban designers, to providing a southern orientation for the house. A structure located on the north side of the dwelling cluster facing south will be the most important one in the complex and is often two stories high. The structures on the east and west sides will be lower in height with openings facing the patio.

Obviously, the preoccupation with orientation is an attempt to maximize the use of solar energy. The traditional house evolved primarily in the northern part of China where winters are harsh. The structural pattern extended south, becoming well-established culturally, and displaying what came to be considered the distinctive features of all Chinese dwellings. The typical house of the well-to-do consisted of a main living/reception room, the master bedroom, a study, and a child's bedroom in the suite on the north; older children's rooms in one of the side suites; those of a married son in the other; and servants' quarters on the south side. In this secluded house, residents enjoyed privacy in the open air of the courtyard. Sometimes the servants had their own courtyard, separated from the family's by means of a dividing wall or a veranda.

The walls of the rooms facing the courtyard contained windows or doors, the windows being of paper despite the wintry wind blasts. The privy was usually built in a corner separated from the house and waste disposal was taken care of by private-enterprise collectors and sold to farms. The bathroom did not exist as such; baths were managed by means of basins or tubs brought to the private rooms by servants.[1]

This prototypical house became commonplace during the Qing dynasty (1644–1911) and is an evolutionary product of socio-economic and environmental forces. Contemporary attempts have been made in modern neighborhoods and housing complexes to continue its traditions. This cultural tendency is strengthened by certain factors in the present socialist regime, such as the neighborhood social order and management, the establishment of a management unit for residency and work, and the combining of the place of work with the place of living.

Vernacular House Design

Much vernacular architecture in China represents the geographical diversity of climate, soil, and available building materials and also reflects variations in cultural and ethnic groups. The vernacular buildings demonstrate the accumulated experience in design, artistic expression, and the adjustment to diverse environmental conditions. The vernacular architecture is represented in dwellings, palaces, temples, and other structures. Unfortunately the least permanent throughout history, these dwellings usually did not survive. Yet they are the fulfillment of functional and aesthetic individual and family needs. The cave-dwelling design concept is not isolated from vernacular housing. As a general background it is worth reviewing the typical above-ground vernacular housing types (fig. 1).

In northern China, the Han ethnic group that actually constitutes the majority of the Chinese people, historically used the design commonly known as the Beijing courtyard house, consisting of a courtyard surrounded by rooms (fig. 1-H). The main building of the complex has a southern orientation and receives the most sunlight. Traditionally, the gate of the house was located at the southeastern corner of the complex. The entrance did not lead directly into the main courtyard but instead

was divided into two parts, an inner gateway sheltered by an overhang, and a reception area giving access to the courtyard. The main structure facing south was used by the head of the family while the eastern and the western buildings were used by the younger family members. All the rooms were connected by corridors. Another building often faced south behind the main building complex with its own long, narrow, rear courtyard. Each of the courtyards was enclosed by rooms and walls, windowless when facing exterior space, completely isolating the house from the outside environment. There are also medium-sized dwellings south of the Yangtze River similar to this traditional Beijing-house style featuring enclosed rectangular courtyards running along an east-west axis (fig. 1-D and 1-G). Often the rear building is more than one story high. Windows on the upper level and patterned perforations in the boundary and divider walls receive the maximum effect of light and ventilation and avoid direct solar radiation in this humid, hot climate.

In the southern part of Fujian province and the northern parts of Guangdong and Guangxi provinces in southeastern China, the Hakka (Ke-jia "guest family") tribes have lived collectively in special structures since the third century (fig. 1-E). There are two forms, one that has a courtyard with a large building in the center flanked by small buildings on the sides; the roofs are at varying heights for the different sections. The other type is a multi-storied building, which may be rectangular or circular in plan, built of brick and earth. The building on the perimeter is usually three or four stories high, and the structure within has one story with several courtyards. The exterior wall, built of rammed earth one meter thick supported by wooden frame construction, has small windows and gives the overall appearance of a fortress. In the sub-tropical regions of southeastern China, such as the provinces of Guangxi, Guangdong, Yunnan, and the Hainan Islands, the climate is humid and hot. The native population elevates the dwellings on stilts to facilitate air circulation (fig. 1-B). The elevated structures of the Dai minority people of Yunnan have steep hip and gable roofs and leave the ground beneath for livestock, storage, and other usage.

Tibetan-style dwellings in Tibet, Qinghai, Gansu and western Sichuan provinces are constructed with stone walls and closely-spaced joists for floors and flat roofs (fig. 1-F). Rural dwellings with no courtyards are built two or three stories high on hillsides. The ground floor is used for livestock; the

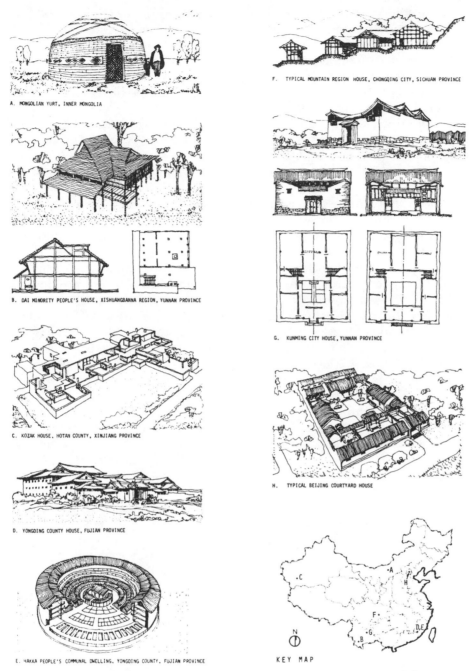

A. MONGOLIAN YURT, INNER MONGOLIA

B. DAI MINORITY PEOPLE'S HOUSE, XISHUANGBANNA REGION, YUNNAN PROVINCE

C. KOZAK HOUSE, HOTAN COUNTY, XINJIANG PROVINCE

D. YONGDING COUNTY HOUSE, FUJIAN PROVINCE

E. HAKKA PEOPLE'S COMMUNAL DWELLING, YONGDING COUNTY, FUJIAN PROVINCE

F. TYPICAL MOUNTAIN REGION HOUSE, CHONGQING CITY, SICHUAN PROVINCE

G. KUNMING CITY HOUSE, YUNNAN PROVINCE

H. TYPICAL BEIJING COURTYARD HOUSE

KEY MAP

Fig. 1. Typical vernacular Chinese houses representing different regions and ethnic groups.

other floor for living quarters. Tibetan urban dwellings are usually two stories high in a formal arrangement enclosing a small courtyard.

Uygur and Kazak people in Xinjiang, where the climate is hot and dry, build courtyard dwellings with closely spaced wooden joists to form flat roofs and adobe to form the exterior walls (fig. 1-C). The front is often colonaded, and the walls are richly decorated with paintings in relief. There are skylights in the ceilings and a minimum number of windows in the side walls.

Nomadic Mongolian and Kazak people in the northernmost part of China use movable yurts in their windy, dusty, arid environment (fig. 1-A). The yurt is made of a bamboo skeleton covered with thick wool felt. The size of a yurt is 4 to 6 meters in diameter and 3 meters in height at the center. At the center top there is an opening for light and ventilation.[2]

The cave dwelling designs described in this book follow some of the basic principles of the vernacular Chinese houses. The concepts of the courtyard and the southern orientation of the important rooms are introduced in a variety of forms. Most of the cave dwellings exist in the warm, dry climate of northern China where abundant loess soil plays a decisive role in the development of these distinctive structures.

According to statistics, there are some areas where cave dwellings represent the majority of shelters. For example, in Qingyang in the eastern part of Gansu province, cave dwellings in loess soil constitute 83 percent of the total number of shelters. The Qingyang and Pingliang areas were researched by the Chinese Investigation Group, which focused on thirteen different locations and examined one hundred cave dwellings and forty garden houses. Many of the cave dwellings in this region were constructed in gullies, 551 exist in the Qingyang area alone, each one measuring as much as 5 kilometers long. Usually the loess soil develops vertical cliffs and many gullies through erosion. Consequently, two types of cave dwellings exist: one in the cliffside and the other on the flat area on top of the plateau.

Because of past overuse of wood in China, farmers, the majority of the population, have been limited in their choice of construction materials. Moreover, the preparation of burned brick consumes coal which also is in limited supply in some regions. The options left are adobe, earth (for rammed earth walls), or below-ground space. The latter seems to be much cheaper, quicker, and easier to use; is expandable; saves energy, among other advantages.

Background of Cave Dwellings

China constitutes approximately 9.6 million square kilometers, about one-fifteenth of the earth's land surface, and is the third largest country in the world geographically.

Chinese cave dwellings are unique not only in their features and design, but are also socio-economically, environmentally, and pedologically unique. They are confined to north and northwestern China where the loess soil (yellow soil) is distributed, ranging from Shanxi and Henan provinces in the east to Gansu province in the northwest (approximately 2,500 kilometers), and from Shaanxi and Ningxia provinces in the north to the Qinling Mountains in the south (approximately 1,000 kilometers). It is an arid and semi-arid zone and both the soil and climate have wide impact on the design and the distribution of these cave dwellings.

The Qinling Mountains that run east-west form a dividing barrier between north and south China, between arid and humid China, between irrigated and dry farming, and between loess soil and other types of soil. They stand in the way of humid air, especially summer rains, penetrating to the north and also prevent the warm dry air of central Asia from penetrating to central and south China. North of the Qinling Mountains are the five provinces (Gansu, Shaanxi, Shanxi, Henan, and the Ningxia Autonomous Region) where loess soil and cave dwellings are to be found.

The general estimate of the Chinese cave-dwelling population is between thirty to forty million people. They constitute 20 to 25 percent of the total population of the five provinces, or 4 percent of the total Chinese population. Their distribution is not even in each province as much depends on the quality of the loess soil, the physiographical and geological formations of the region, and to a great extent historical events such as regional conflicts and wars. In any case, in Henan province, they are primarily located in its north and northwest; in Shanxi province, in its north, west, and southwest; in Shaanxi province, in its north and southwest; in Gansu province, in its east in the Qingyang region; and in the Ningxia Autonomous Region in its south.

Historically this cave dwelling region is the cradle of the Chinese civilization. When Neolithic man moved from nomadic and semi-nomadic tribal conditions to more socially cohesive tribal villages, his dwellings were located below ground or partly below ground and served to accommodate both humans and domesticated animals. One good example of this pit shelter of the Neolithic period is the village of Banpo or Pan-po-ts'un (at Xi'an City), which is comprised of largely semi-subterranean houses surrounded by a deep ditch for protection. Many such villages are also found in the Chung Yuan area around the Wei He River in Shaanxi province and were the center of the Yangshao Neolithic culture that is considered to be the earliest in the Yellow River valley.

The region is crossed from west to east by the gigantic meandering Yellow River. As with many other early civilizations (e.g. Mesopotamia, Indus Valley, Egypt, Central America), the Chinese also started in dry regions and grew on irrigated agriculture. Paleolithic as well as Neolithic civilizations started primarily in Henan, Shanxi, and Shaanxi provinces. Archeological findings in this area indicate that man by 3000 B.C. had already used the loess soil for the development of cave dwellings for habitation.

Cave dwellings continue through the centuries, especially during the Warring States Period (475–221 B.C.), which was a transitional phase from slave to feudal society. This period had a progressive significance ideologically, which brought social changes and introduced conflicts between the Confucianist and Taoist schools, and ended when Qin Shi Huang (first emperor of the Qin dynasty) unified the first feudal state in China. In any case, the wars and conflicts created refugees and homeless people on a large scale and cave dwellings served as an optimal solution for their habitation. Military organizations also required shelter on short notice for soldiers, and again cave dwellings were used. For example, at the end of the Sui dynasty (A.D. 581–618) the Emperor Li Mi stationed 300,000 troops north of Luoyang City, Henan Province, in preparation for a decisive battle. Digging cave dwellings provided the necessary shelters in a short time without requiring expenditure of additional funds for building materials. Similarly in this century, Mao Tse-tung terminated the Long March of his Red Army in the Yan'an City region, Shaanxi province, as the cave dwellings of the region were available for shelter and provided excellent protection against Japanese air raids. Thus, according to historians, Chinese cave dwellings have been continuously in use up to our time.

Similar to other Chinese vernacular dwellings, the cave dwellings were also influenced by environmental constraints of soil, climate, terrain, and physical forms. Those factors have shaped the configuration of cave dwellings and the farmers who generated them in order to maximize their own survival were always concerned with living in harmony with their natural resources. There are also full villages which have been constructed below ground (fig. 2).

For the Chinese farmer, the cave dwelling suited

Fig. 2. Bird's-eye view of a cave-dwelling village east of Lanzhou City, Gansu province.

his socioeconomic and historically chaotic political condition. The advantages to his standard of living are obvious. He was warm in winter and cool in summer (thus energy-saving); required no building materials (China has a perpetual shortage of wood); and needed only a simple technology using the readily available axe, shovel, and basket. All in all, a cave dwelling consumes a minimum of land, leaving the surface for agricultural use, and saves the scarce resources available to the Chinese farmer.

Physiographically the region forms a plateau made of loess soil and is bounded by several high mountain ranges on the east, south, and west, leaving the hilly north open to central Asia (fig. 3). The highest mountains are the Qilian Mountains in the west (4,000 meters), followed by the Qinling Mountains in the south (3,000 meters), and the Taihang Mountains in the east (2,000 meters). To some extent there is a physiographical unity in this region, which also makes the region climatologically and pedologically distinctive. The loess soil strip extends from east to west and covers an area of more than 600,000 square kilometers, approximately 6.6 percent of the country. This is the largest loess soil concentration in the world. The altitude of this region is between 200–2,400 meters, the depth of

the soil is between 50 and 300 meters, and it is crossed by the Yellow River along 2,000 kilometers of its course. The loess soil accumulation in this region is primarily a result of eolian movement of dry dust carried by deflation from central Asia, though there are some spots of loess soil that are a result of fluvial processes. This dry wind still blows from the Gobi Desert, Outer and Inner Mongolia, during most seasons of the year, carrying fine particles of sand and soil.

The loess soil was formed in the Pleistocene epoch of the Quaternary period, beginning approximately one million years ago and continuing to the present. It is composed of salty minerals from loamy sands rich in nitrogen, phosphorus, and potassium, which are good for agriculture. The lack of vegetation cover, the dryness, and the looseness of the soil have intensified erosion of the loess and created its distinctive forms and landscape. This erosion process occurs primarily at the edges of the plateaus, where many gullies and gorges form. More distinctively, the erosion process is vertical rather than horizontal or diagonal, as occurs in other types of soil. Consequently it is common to find land forms such as sink holes, cliffs, tablelands, elongated loess mounds, round loess mounds, land-

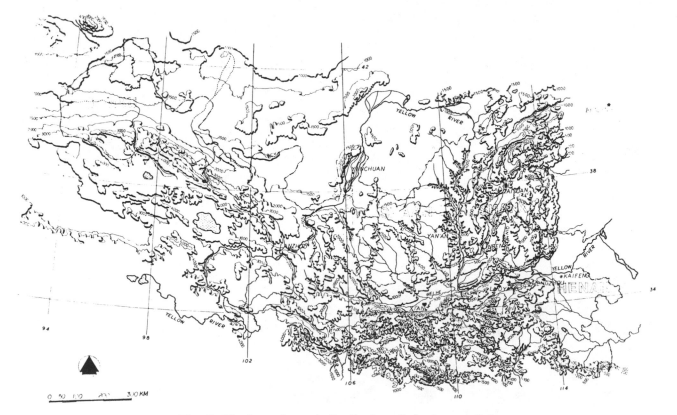

Fig. 3. Physiography and distribution of the loess soil in northern China.

slides, columns, vertical cracks, gullies and ravines.

The soil's density increases with the depth of its geological Quaternary stratification. Most of the cave dwellings are located within the less dense Q4 and Q3 levels, while Q2 and Q1 levels are not commonly exposed. Thus the design of cave dwellings has required much consideration to avoid or minimize collapse in case of earthquake vibrations or intensive erosion from rainfall. In most cases the loess soil is uniform with stratification free from stone, and is composed of fine particles that easily support a straight cut and the creation of a wall. When the soil has little or no moisture, it is very solid and forms a hard crust, thus the serious threat for the cave dwelling is from high moisture and rain which may intensify funnels and cavities and result in collapse. This soil also requires a special agricultural technique for cultivation and maintenance.

Climatically the region ranges from dry to semi-dry conditions. This climate determines the level of the water table, the degree of moisture in the soil, and the nature of the rain pattern in the region. There is, however, a variation in aridity within the region itself, the most dry area being the northwest, and the most humid area being the southeast. The majority of the rainfall occurs from July to September and may range from less than 100 millimeters in the northwest to approximately 700 millimeters in the southeast, with snowfalls in small quantities during the winter in some parts of the region (fig. 4). The rain pattern is characterized by scarcity, storminess, torrentiality, short duration, and a high ratio of run-off, which, with the lack of vegetation cover, intensifies erosion. The combination of heat and humidity during the summer introduces serious problems in cave dwellings, with a consequential high level of relative humidity and moisture in the caves.

Basically the region has two lengthy seasons, summer and winter, with a brief autumn and spring. The temperatures range from -18 C to 30 C, with the highest annual average occurring in the south, and the lowest annual average occurring in the north (fig. 5). Wind velocity differs from one part of the region to another. In general, high ve-

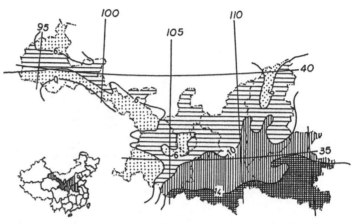

Fig. 5. Average annual Celsius temperature in the five provinces.

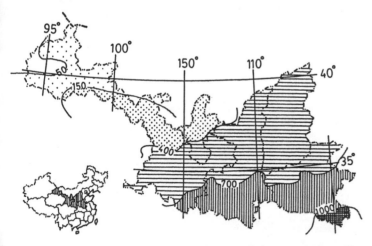

Fig. 4. Average annual precipitation of the region in millimeters. Rainfall occurs mainly during the months of July and August. A small amount of snow falls during January and February.

locity exists in the northwestern part throughout the year. However, the constant pattern of velocity is in the north (supporting the theory that loess soil accumulation is an eolian process). January is the coldest month of the year for the whole region, and July and August are the warmest months of the year. Frost days throughout the year are common. In the lowland parts, more than one hundred frost days may take place, and seventy-five or more frost days may occur in the major part of the region. This is a stressful climate in which differentiation between diurnal and seasonal temperatures is extreme. With such climatic conditions, the cave dwelling proves to be a uniquely adapted structure.

In general, it can be stated that the cave dwelling region of China has a continental climate primarily influenced by central Asia. This climate influences the land form, and this in turn shapes site selection for the cave dwellings. Finally, the loess soil has an impact on the construction of the cave dwellings. It requires simple technology, uses local materials, provides low cost construction, and lastly it has good thermal performance.

A. PIT CAVE DWELLING

B. CLIFF CAVE DWELLING

C. EARTH-SHELTERED (ABOVE GROUND) DWELLING CONSTRUCTED OF ADOBE OR BRICKS, SIMILAR IN DESIGN TO CAVE DWELLING

D. SEMI-BELOW-GROUND DWELLING ON FLAT SITE

E. COMBINED BELOW- AND ABOVE-GROUND DWELLINGS, CLIFF SITE

Fig. 6. Different types of earth-sheltered dwellings in northern China.

Cave-Dwelling Design

Cave dwelling designs have evolved throughout the history of China, and have been influenced by environmental and socio-economic factors. Although there are many types and associated forms of cave dwellings, two basic designs have emerged: the cliffside dwelling and the pit cave dwelling (fig. 6).

CLIFF CAVE DWELLINGS

The cliff cave dwellings are primarily dug into the vertical face of terraced cliffs, with the excavated soil spread to build a terrace and create a courtyard (fig. 6-B). This type of dwelling consumes a minimum amount of land, and is located in terrain not usually suitable for agriculture. The dwellings can be extended one horizontal linear level or more to form a terraced cave dwelling settlement (fig. 7). This form is found in the gullies and gorges of the region, or at the edges of the plateaus.

Cliff-dwelling sites are usually chosen with a preference for a southern orientation, availing the dwelling of optimal exposure to natural light and minimal exposure to the prevailing wind patterns. Although cliff types do not normally have a chimney or any other ventilation system, they are better ventilated than the pit type. A variety of structural forms resulting from their adjustment to the topographical configurations have been introduced. Essentially there are one-cliffside, two-cliffside, and three-cliffside cave dwellings in which the remaining sides are enclosed by a wall to form a courtyard though there are other forms diverging from these patterns. Cliff dwellers enjoy direct eye contact with the surrounding area, particularly to the bordering lowlands, overcoming the claustrophobic effect of confined below-ground spaces, they also receive more light and sunshine, in contrast to the pit type cave inhabitants.

Overall the cliff cave dwellings are more advantageous to human habitation and introduce greater architectural variations than the pit cave dwellings (fig. 6-E). Many of the cliff dwellings are integrated

Fig. 7. Cliff cave dwellings, earth-sheltered habitats, and above-ground structures in the eastern part of Yan'an City. More than 60 percent of the structures are of the first two types.

with above-ground structures, and in other cases we find above-ground structures as independent units within the courtyard of cliff dwellings, or combined with the cliff unit as a transitional portion (fig. 8).

Fig. 8. An integration of above-ground (in front) and below-ground dwellings in a terraced cliff in Xi Cun Township, Gong Xian County, Henan province.

PIT CAVE DWELLINGS

The pit cave dwellings are primarily located on flat or rolling terrain, and their design is influenced by such an environment. The overall design plan consists of a large pit, generally 8 by 10 meters, and 7 to 9 meters in depth (fig. 9). Usually two rooms

Fig. 9. Pit cave dwelling courtyard opening, Zhong Tou Village, near Luoyang City.

are dug on each of the four sides of the pit on the level of the courtyard. They are generally 3 meters wide and 7 or more meters in length, and designed

in vault form, 2.5 or more meters in height (fig. 6-A). This type consumes more land than cliff type dwellings. Because of the risk of water penetration of the soil and resultant collapse, the soil above the complex is not generally used for agriculture, removing even more land area from agricultural production. Yet the pit dwellings provide the traditional Chinese concept of enclosure and intimacy within the dwelling unit. Some introduce two courtyards within the complex following the Chinese vernacular concept of a transitional courtyard, with the entrance forming a graded stairway constructed partly underground and partly open to the sky. One complex may consist of eight rooms or more, those facing south being used as living space for senior members of the family, and those facing north usually being used for stabling animals and for storage.

EARTH-SHELTERED HABITATS

Another kind of dwelling commonly used by the rural Chinese are different types of earth-sheltered habitats (fig. 6-C). These are above-ground structures built of vaulted, linearly attached rooms constructed of stone or bricks with thick walls approximately one meter wide and with heavy roofs. In many cases such a structure is also attached to the cliff (fig. 10). Their thermal perform-

Fig. 10. An earth-sheltered dwelling attached to a cliff, Gao Me Wan Village, east of Yan'an City, Shaanxi province. Note the earth cover above the structure and the narrow walls between units.

ance is similar, but not identical, to that of the cave dwellings; this is achieved because of the thickness of the walls, which introduces a seasonal thermal time lag. Although commonly constructed above ground, there are some that are built in a semi-subterranean form, integrated with the cliffside or with the surface of the ground (Fig. 6-D). Construction of earth-sheltered habitats is on the increase not only within the cave dwelling region but throughout China today. Their overall design resembles the cliff cave dwellings.

FORM

The dominant form of one room unit of the cave dwellings, whether cliff or pit type, is the vault. Although most of the commonly used vaults are the parabolic type there are other types such as the pointed horseshoe, semicircular, elliptical, and various others. The advantage of the vault form is its extendability. However, its width is limited usually to 3 or 3.5 meters because of soil constraints, yet its height can exceed this measure to 4 or more meters. In some cases, the ceiling is slanted toward the back of the unit, thus concentrating and intensifying the light emanating through the higher frontal area. The entrance to the unit is through this facade, which has one door and usually one large window in the middle, and another smaller window at the top for ventilation. The overall amount of light entering the room through the front is limited, especially at the very inner part of the room. Ventilation, light, and high summer humidity are the major problems of the cave dwellings.

In most cases the vault is made of a simple cut through the soil and the wall is mortared by loess soil mixed with straw. The facade of the room at its cliff face is covered by stone or by bricks, which stabilizes the wall and avoids landslides. Recently, with the new economic reforms, and the improvement in the standard of living, the cave dwellers have been covering the vault face with an aesthetically pleasing light-colored burnt brick, and gaining a corresponding increase in the structural strength and an elimination of dust falling from the walls.

Thermal Performance

Studying, analyzing, and understanding the vernacular experience of soil behavior is of extreme importance for our future earth-sheltered space. There are two basic rules that need to be understood concerning thermal behavior within the soil and the advantages that the soil offers in its turn. These rules are not fully comprehended by many architects, consequently their modern designs lack

that efficiency and optimality that such a study can offer. These two rules are: (1) the soil is an efficient thermal insulator diurnally, yet (2) its primary significance is as a seasonal heat retainer.

Diurnal soil insulation occurs in a twenty-four hour cycle. Solar radiation affects the soil throughout the day to a very limited depth ranging between 7 and 10 centimeters. This depth depends on soil composition, the intensity of the heat, the degree of moisture, and the amount of vegetation cover (i.e. grass, shrubs, trees). The greatest heat gain occurs primarily between 2:00 and 4:00 P.M. On the other hand, the soil dissipates the heat during the night. Heat dissipation depends primarily on the degree of relative humidity of the air. In dry air, such as that of desert and semi-desert areas, heat dissipation will take place quickly, mostly during the first few hours of the evening. However, in a humid region, the relative humidity will retain the heat of the air, and consequently soil heat dissipation will be very slow. In both cases, maximum soil heat dissipation will take place toward the end of the second part of the night. This is the reason frost occurs quite often in a desert area when temperatures go below the freezing point toward dawn. Therefore space below the surface of the soil is insulated from the diurnal intensity of solar radiation. The deeper the space, the higher the insulation. Along with this process there is the slow diurnal movement of heat deep into the earth that transfers a fraction of the day's heat.

As designers of earth-sheltered space, seasonal soil heat retention is our main concern. The daily heat movement is a continuous process throughout the season and the heat will be retained for use during the following season. This process usually takes place down to a depth of 10 to 11 meters below the surface. At that depth the soil temperature remains relatively stable with only a little fluctuation from one season to another. In other words, the closer to the surface, the more the seasonal temperature graph fluctuates between the maximum and the minimum as it follows the seasonal air temperature fluctuation. In short, the mass of soil functions as a seasonal heat retainer. Thus, the summer's high air temperatures will be retained by the soil for use in the winter and the winter's low air temperatures will be retained for summer use.

An understanding of the above two processes is essential for our design elaboration of below-ground space. However, it is not necessary to design the space ten meters deep to achieve a stable warm temperature in the winter and a cool environment in the summer. We can obtain similar results at a shallower depth by covering the soil with vegetation (grass or trees), by watering at night (the warmed soil will cause the heat to penetrate deeper), by building above ground, or in special circumstances, by a special study. We can conclude that there are several ways of diverting the natural system to achieve different results.

The question arises, can we stay above ground and obtain similar results? The obvious answer is yes. Ancient civilizations' mostly vernacular builders understood this process through experience, observation, and intuition. In the Mediterranean world, considering that the climate is defined as warm and dry in the summer and cool and wet in the winter, diurnal summer temperatures are extreme because of the aridity and semi-aridity; seasonal temperatures fluctuate widely and dry air is dominant most of the year. Therefore, structures are enveloped by walls 1 to 2 meters thick. This author's study of heat gain and heat loss in the vernacular housing that dominated Jerusalem until the 1930s can be summarized as follows:

1. The walls of the dwellings were between 1 to 2 meters thick and consisted of an interior wall made of small stones, 5 to 10 centimeters thick, mortared and whitewashed with burned limestone, and an exterior wall of large stones, 20 to 30 centimeters thick. The space between the two walls was filled with dirt and gravel. This massive wall was required for structural purposes to support the second floor, for protection, and most of all for climatic reasons (fig. 11).

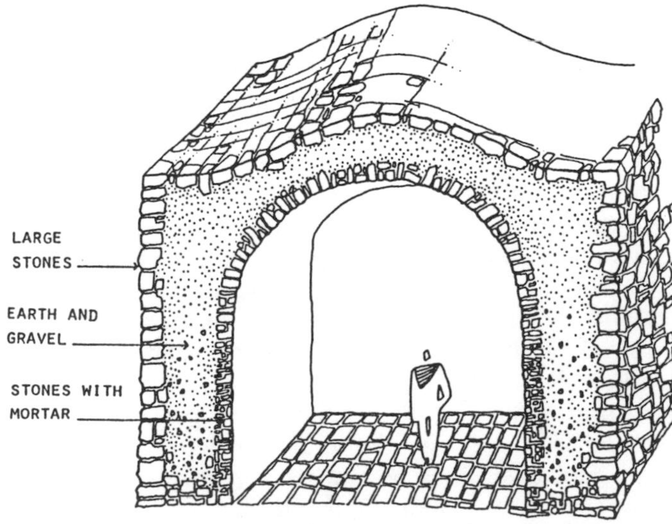

2. The hot and dry climate of the summer strikes the outer wall and consequently a wave of heat progresses toward the inner wall. This heat is retained and reaches the interior wall by the beginning of winter. Thus the interior wall will be radiating the heat toward the interior space and warming it during the winter when the heat is most desired. The thickness of the wall and its soil composition will determine the duration of this heat supply during any given season. The supply may last two or three months of the winter. On the other hand, a similar process occurs in the winter when the outer wall absorbs both low temperatures and the high relative humidity of the rain and each wave begins to move toward the interior wall. The humidity will move by a capillary system and both waves will arrive at the interior wall by the end of April or early May, the beginning of warm weather. A supply of low temperature combined with high relative humidity will be provided to the indoor space as compared to the outdoor space. Consequently, the indoor ambient temperature will be very comfortable and pleasing during the early part of the summer season and some of the later part as well. The duration of the coolness in the summer will depend on the thickness of the wall as well as the intensity of the cold temperatures which had prevailed during the winter.

The second example of vernacular dwelling design determined by heat gain and loss is the Eskimo winter shelter, the igloo (fig. 12). The igloo is a

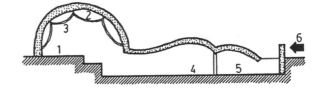

| 1. LIVING SPACE | 2. AIR SPACE | 3. SKIN |
| 4. CURVED CHAMBERED TUNNEL | 5. ENTRY | 6. PROTECTION FROM WIND |

Fig. 12. **General sketch of the Eskimo winter shelter.**

semi-subterranean space, enveloped and covered by a dome composed of blocks of ice. One approaches this space via a descending entranceway. A raised platform in the interior is used as a sleeping bed; the ceilings are covered with animal hides. The ceiling has a small vent hole and a candle or oil lamp provides light and some warmth. Heat, however, is obtained from that which is stored in the

LARGE
STONES

EARTH AND
GRAVEL

STONES WITH
MORTAR

Fig. 11. **General sketch of the traditional Jerusalem house.**

soil from the summer season and from that radiating from the human occupants. The Eskio summer house is similar to the igloo in design, being constructed of wood partly below ground.

The Jerusalem house and the Eskimo igloo are two types of dwellings developed in stressful climates. The majority of the below-ground dwellings constructed throughout history, such as those in China, in southern Tunisia, and in Cappadocia, central Turkey, are in hot and dry or cold and dry climates, which are considered to be stressful. Our study of the southern Tunisia case enforces our conviction that optimum results can be achieved with below-ground dwellings in a stressful climate where temperature and aridity are extreme.

Cave dwellings, due to the thick mass of earth surrounding them, have the ability to retain heat with the least possible heat exchange. Also, wind penetration is minimal. The greatest differentiation between outdoor and indoor temperatures occurs in the afternoon hours, and late at night toward dawn. The outdoor temperature fluctuates throughout the day and the season while the indoor temperature remains relatively stable. Thus the cave dwellings' achievement in thermal performance is seen when comparing indoor with outdoor conditions. It shows the superiority of the former and the inferiority of the latter most of the time. The only exception is the courtyard about 2:00 to 4:00 P.M. on summer afternoons. Our research findings in the twelve Chinese pilot dwellings detailed in this book indicates that on most summer afternoons and especially on rainy days, the courtyard is less comfortable than the outdoor site due to lack of ventilation and to the increase in humidity.

RELATIVE HUMIDITY

Within the cave dwellings, relative humidity is comfortably low in winter. In the summer, relative humidity is higher than that of the outdoors by 9 to 13 percent.[3]

As compared with the outdoors, the temperature within the cave dwellings is higher in the winter and lower in the summer. In the winter the incoming outside air will have lower relative humidity that combines with the higher indoor temperatures. In the summer, the case is the opposite, with the air entering the cave from the outside having higher relative humidity while the cave dwelling itself will have a lower temperature. Condensation will occur when the air temperature decreases to a certain level, thus adding to the high perspiration of the dwelling. This is the main reason

for the greater degree of dampness inside the caves in the summer.[4]

Most Chinese cave dwelling units are designed and constructed in a rectangular form (usually 3.5 by 6 meters by 3.5 meters high). Our summer and winter temperature measurements were always taken at a permanent point close to the center and slightly toward the end of the room on the assumption that such a site would represent closely the average temperature of the dwelling unit. In reality we can assume that there is some differentiation in the dry-bulb and wet-bulb temperatures in different sections of the dwelling.

Horizontally, the front section of the room, close to the door and the windows, will be less humid, brighter, and higher in temperature in summer and lower in the winter. In general, the rear part is expected to be more humid and darker, cooler in summer and warmer in winter. Vertically, some temperature differentiation can also be found between the lower and the upper parts of the dwelling, especially when it is inhabited. In general, we can state that the upper part will be warmer. Thus, construction of a second semi-floor to be used as a bedroom is desirable if the cave is sufficiently high.

The most serious problem with thermal performance is the high humidity at the back of the cave in the summer, especially in the eastern loess soil zone where summer precipitation is rather high: Henan and Shanxi provinces. Added to this is the vapor produced by the inhabitants. In some dwellings the interior walls of the units are covered with moisture. Under such conditions ventilation is essential. Air circulation causes evaporation and reduces the humidity.

The Japanese group that researched the subject of the Chinese cave dwellings found that there is a noticeable difference in temperature of wall surfaces of the structures. When the ceiling temperature was 16.7 degrees C, the side wall temperature was 15.2 degrees C, the back wall was 14.8 degrees C, and the floor temperature was 14.7 degrees C. Also the ceiling temperature was higher by 1 degree C than the air temperature in the center of the cave unit. The floor and the rear wall temperatures were 1 to 2 degrees C lower than the air temperature, and the side wall temperature was about 1 degree C lower than the air temperature in a large number of cave dwellings in the spring. They also found that the vertical distribution of temperature within the cave dwellings differed throughout the day. In winter, and similarly in spring, temperature differences between the upper and the lower space were small in the morning but this difference increased to 4 degrees C by 3:00 P.M. as the outdoor air temperature rose during the daytime.[5]

It was also found that in the spring the relative humidity in the cave dwellings was usually about 20 percent higher than outdoors and the relative humidity at night was as high as 85 to 95 percent.[6] Such high humidity at night was dominant in the cave dwellings in the rainy summer season as well. The lack of ventilation in the Chinese cave dwellings supports this high humidity.

SOIL MOISTURE

An experiment determined that the moisture content of the soil in a new cave was over 12 percent, and in an old cave, less than 5 percent.[7]

To experimentally reduce moisture in the soil,

bricks lining the interior walls of an earth-sheltered dwelling were fired. The outside height from ground to surface was 4.7 meters and the inside height from floor to ceiling was 3.3 meters. The cave was filled with bricks and was fired just as though in a brick kiln. The firing was done gradually over a six or seven day period. The experiment was considered a success and this earth-sheltered cave dwelling now has sintered interior walls, better and more stable than loess or adobe. The compressive strength of the loess in Lishi, Shanxi province, where the experiment was run, is 15 to 19 kilograms per square centimeter. The compression of the adobe (sun-burned bricks) is 4 to 6 kilograms per square centimeter, the brick strength is 9 to 17.8 kilograms, the depth of the sinter was 8 to 12 centimeters, and its moisture content was 3 percent. It was also found that in times of heavy rain,

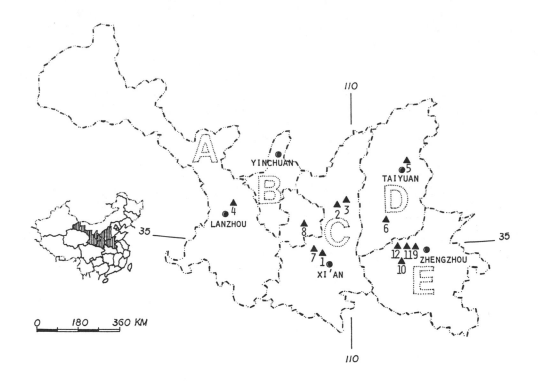

KEY

A. GANSU PRO.
B. NINGXIA PRO.
C. SHAANXI PRO.
D. SHANXI PRO.
E. HENAN PRO.

1. GAO JAI TEAM VILLAGE
2. WANG JIA TERRACE
3. GAO ME WAN VILLAGE
4. YA CHUAN VILLAGE
5. QING LONG VILLAGE
6. DAYANG VILLAGE

7. SHIMA DAO VILLAGE
8. XIFENG ZHEN TOWN
9. BEI TAI VILLAGE
10. XI CUN VILLAGE
11. GONG XIAN TOWN
12. ZHONG TOU VILLAGE

Fig. 13. Location of the twelve cave dwellings researched in four provinces of the loess soil zone.

the humidity in the experimental earth-sheltered dwelling could reach 87 percent. Yet, people feel less damp there than in a regular brick cave, and they do not need to dry their quilts out in the sun or air. They need only open the door for a short time when the sun is shining and the humidity will drop 10 to 15 percent and the walls will not be damp as in other caves. Moreover, according to Jia Kunnan, the sintered earth-sheltered cave dwelling is no more expensive than the standard brick earth-sheltered cave dwelling.[8]

Cases Studied

The focus of this book is the specifics of the design and thermal performance of the twelve cave dwellings surveyed and researched in four provinces of the loess soil region. Six of these are cliff type, and the other six are pit type cave dwellings.

Although there is much uniformity in the loess soil as a whole, there are variations that have influenced the design of the dwelling units and also may have influenced the thermal performance of the soil to some extent. The dwellings have been selected from different geographical regions in the loess soil zone. Included are those located in villages, towns, and cities, two to four from each of the provinces of Shaanxi, Shanxi, Henan, and Gansu (fig. 13). Each example was selected from among a variety of cave dwellings and represents a prototype for its own region. Therefore it represents the climate and, to some extent, the subculture of that region as well (table 1).

The field research activity for each unit included surveying and mapping the entire cave dwelling complex, its design, and its architecture; interviewing the tenants; extensive photographing of the complexes; and, last but not least, measuring the indoor and outdoor dry-bulb and wet-bulb temperatures within different sections of the units every hour for a twenty-four hour period in summer and winter. All the units, whether pit type or cliff type, except Dwelling No. 3, are cut deeply below the surface of the soil and represent the common diversified experience of the Chinese cave dwellings. Dwelling No. 3, a cliff cave dwelling, is remarkable because it has one wing buried deep in the cliffside soil and has earth-sheltered units for the other wing. The latter are structures made of stone and covered with earth where the roofs can be used for agricultural production. Obviously, these two different types constructed adjacent to each

Table 1. Cave Dwellings Researched in the Four Provinces

PROVINCE	COUNTY/REGION	SETTLEMENT	DWELLING/ FAMILY
1. Shaanxi	Liquan County	Gao Jia Team Village	Gao Ke Xi
2. Shaanxi	Yan'an region	Yan'an (west), Wang Jia Terrace	Gou Shengzhi
3. Shaanxi	Yan'an region	Yan'an (east), Gao Me Wan Village	Zhang Yen Fu
4. Gansu	Gao Lan County	Ya Chuan Village (near Lanzhou)	Cao Yiren
5. Shanxi	Yang Qu County (Taiyuan City)	Qing Long Village	Zhao Qingyu
6. Shanxi	Linfen City	Dayang Village	Cui Mingxing
7. Shaanxi	Qian Xian County	Shima Dao Village	Bai Lesheng
8. Gansu	Qing Yang region	Xifengzhen	Xing Xigeng
9. Henan	Xing Yang County	Bei Tai (near Zhengzhou City)	Tian Lu
10. Henan	Xi Cun Township, Gong Xian County	Xi Cun Village 15th Team	Yin Xin Yin
11. Henan	Gong Xian County	Gong Xian Town	Li Songbin
12. Henan	Mang Shan Township of Luoyang City	Zhong Tou Village (near Luoyang City) No. 199 in Shang To Village	Liu Xueshi

other produced different thermal performances.

The researched cave dwellings include six of the cliff type and six of the pit type (fig. 14). Each represents a unique example of design and thermal performance. Together, these twelve dwellings represent the prototypes of Chinese cave dwellings of the loess soil region.

The Chinese house in general, and the cave dwellings in particular, differ from Western culture in functional division of room space. The Chinese habitation tends to mix the functions of the home units. Often we may find that one room of a cave dwelling is used as a kitchen as well as a bedroom, and the other rooms mix sleeping quarters with storage. In general, there is no guest room and not what is commonly known as a living room in the cave dwellings. This condition exists even in cave dwelling complexes that have large numbers of rooms and are inhabited by small families.

The overall design of the cliff type and the pit type dwellings is similar. A basic concept of the Chinese house design, as well as of the cave dwelling complex, is the centralized courtyard surrounded by a set of rooms. The development of a second floor in the cave dwellings is uncommon in

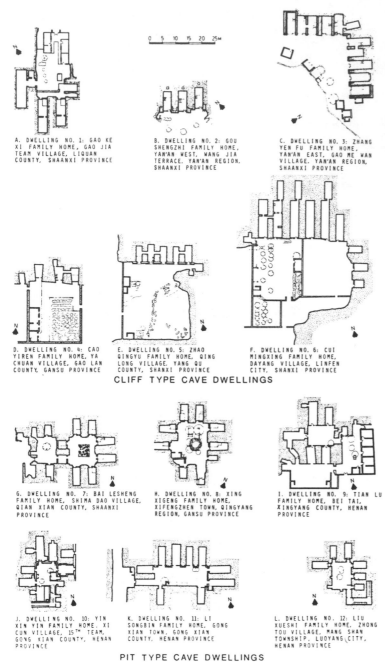

A. DWELLING NO. 1: GAO KE XI FAMILY HOME, GAO JIA TEAM VILLAGE, LIQUAN COUNTY, SHAANXI PROVINCE

B. DWELLING NO. 2: GOU SHENGZHI FAMILY HOME, YAN'AN WEST, WANG JIA TERRACE, YAN'AN REGION, SHAANXI PROVINCE

C. DWELLING NO. 3: ZHANG YEN FU FAMILY HOME, YAN'AN EAST, GAO ME WAN VILLAGE, YAN'AN REGION, SHAANXI PROVINCE

D. DWELLING NO. 4: CAO YIREN FAMILY HOME, YA CHUAN VILLAGE, GAO LAN COUNTY, GANSU PROVINCE

E. DWELLING NO. 5: ZHAO QINGYU FAMILY HOME, QING LONG VILLAGE, YANG QU COUNTY, SHANXI PROVINCE

F. DWELLING NO. 6: CUI MINGXING FAMILY HOME, DAYANG VILLAGE, LINFEN CITY, SHANXI PROVINCE

CLIFF TYPE CAVE DWELLINGS

G. DWELLING NO. 7: BAI LESHENG FAMILY HOME, SHIMA DAO VILLAGE, QIAN XIAN COUNTY, SHAANXI PROVINCE

H. DWELLING NO. 8: XING XIGENG FAMILY HOME, XIFENGZHEN TOWN, QINGYANG REGION, GANSU PROVINCE

I. DWELLING NO. 9: TIAN LU FAMILY HOME, BEI TAI, XINGYANG COUNTY, HENAN PROVINCE

J. DWELLING NO. 10: YIN XIN YIN FAMILY HOME, XI CUN VILLAGE, 15TH TEAM, GONG XIAN COUNTY, HENAN PROVINCE

K. DWELLING NO. 11: LI SONGBIN FAMILY HOME, GONG XIAN TOWN, GONG XIAN COUNTY, HENAN PROVINCE

L. DWELLING NO. 12: LIU XUESHI FAMILY HOME, ZHONG TOU VILLAGE, MANG SHAN TOWNSHIP, LUOYANG CITY, HENAN PROVINCE

PIT TYPE CAVE DWELLINGS

Fig. 14. Overall plans of the twelve cave dwellings researched.

China because of the relatively weak stability of the loess soil. Yet some dwellings have developed this second floor on a small scale for storage and for hiding in troubled times. Moreover we have found that some cave dwellings have another space below the cave floor that in some cases provides an escape route. Although in the past, the dwellers of one cave complex were one single family, we can find today that such a complex is usually shared by more than one family.

Although improvements have taken place recently in both types of cave dwellings by adding bricks to the facade and inside at the vault, the design has not changed for many centuries.

It becomes noticeable that the cliff type dwellings have more advantages than the pit type dwellings. In a country like China, where arable land is scarce and essential for the farming communities, the cliff type consumes much less land than the pit type.

Citing table 2, if we define the lot line of each of our twelve cave dwellings surveyed, we find that the average dimensions of the cliff type are smaller (29 by 22 meters) than the average of the pit type (36 by 31 meters). The total average of the cliff type lot is around 50 percent (638 square meters) of the average size of the pit type lot (1,116 square meters), yet the average number of residents in the cliff type is lower (6 people) than that of the pit type (8.5 people). Similarly, the density per square meter, with the average area per resident in the cliff type (98 square meters per resident) is smaller than the pit type (138 square meters per resident). The range of the lot size also differs in each type, with the largest lot size among the six cliff types being 924 square meters, compared with the pit type's largest lot size of 1,344 square meters. The smallest lot size among the cliff types is 81 square meters, compared with the pit type's smallest lot size of 783 square meters. In addition, the cliff type, especially those among the high cliffs, do not utilize land suitable for agriculture by virtue of their location. Their courtyards are created from soil dug from the cliff to create the room units and to terrace the cliff face, not from soil otherwise usable for agriculture, whereas the pit type courtyards are usually dug out on flat or rolling areas suitable for agriculture.

In general, the cliff type enjoy a larger courtyard than the pit type (table 3), with the average cliff type courtyard (381 square meters) larger than the average pit type courtyard (115.71 square meters), the largest cliff type courtyard (728 square meters) larger than the largest pit type courtyard (319 square meters), and the smallest cliff type courtyard (85 square meters) larger than the smallest pit type (58 square meters). Additionally, the cliff type courtyards have more above-ground structures constructed within than the pit type courtyards. The total volume of earth to be dug from the cliff type, excluding the courtyard, is larger than that of the pit type, with the average volume of soil dug from the cliff face and below ground rooms (1,698 cubic meters) greater than the average volume for the pit type (1,365 cubic meters). This is because the pit type courtyard has to be dug completely below

Table 2. Lot Size* of the Twelve Dwellings Surveyed and Land Consumed by Cliff and by Pit Cave Dwelling Units

PROVINCE	DWELLING NUMBER	DWELLING TYPE	LOT SIZE			
			DIMENSIONS (meters)	(meters²)	NUMBER OF RESIDENTS	AREA PER RESIDENT (meters²)
			(1)**	(2)	(3)	(4)(2-3)
Shaanxi	1	Cliff	24 × 13	312	5	62
Shaanxi	2	Cliff	18 × 4.5	81	5	16
Shaanxi	3	Cliff and Earth-Sheltered	39 × 45 (Irregular shape)	464 ***	10	46
Gansu	4	Cliff	26 × 20	520	5	104
Shanxi	5	Cliff	33 × 28	924	4	231
Shanxi	6	Cliff	35 × 25	875	7	125
Shaanxi	7	Pit	42 × 32	1,344	11	122
Gansu	8	Pit	36 × 31.5	1,134	8	141
Henan	9	Pit	37 × 35	1,295	11	117
Henan	10	Pit	32 × 31	992	9	110
Henan	11	Pit	41 × 28	1,148	8	143
Henan	12	Pit	29 × 27	783	4	195
Average			32 × 26	822	7.2	114
Average, Cliff (Nos. 1–6)			29 × 22	638	6	98
Average, Pit (Nos. 7–12)			36 × 31	1,116	8.5	138

*For Pit dwellings, lot dimensions are that part of the "cover" within which is included all the below-ground space units, plus three meters width added to four sides. For cliff dwellings, lot size includes the terraced patio and the "cover" of the below-ground rooms.

**Number above each column is given for identification of the column, especially when it is reused in the following tables.

***Requires special calculation because of irregular shape. Dimensions are given for the two longest walls.

ground with dimensions of approximately 10 by 10 by 10 meters, while the cliff type courtyard is created on a terrace cut from the diagonal slope creating a cliff face and enlarged terrace. It is also the result of the larger and deeper rooms of the cliff type, which require extensive excavation of soil, which is added to the terrace. In general, we can state that the pit type courtyard uses more labor for digging and carrying the soil away than the cliff type courtyard construction, which utilizes the soil on the site.

Table 3 shows a clear distinction between the cliff type and pit type dwellings. In the six cliff type dwellings, we found the average number of rooms is 7.3, compared with 9.3 for the pit type. The combined average is 8.3, but the cliff type is compensated in its average below-ground room shortage by above-ground structures. The average number of above-ground structures in the cliff type complex was 3.3, while that of the pit type complex was 1.5. Thus, if we combine the below-ground rooms with the above-ground rooms the average number is almost even in the two types of dwellings.

The average floor area of below-ground covered space is higher in the cliff type than in the pit type (508 square meters versus 309 square meters); in

the bedrooms, storage rooms, mixed function rooms, and other rooms (except the kitchen), we found that the cliff type dwelling room dimensions averaged greater than the averages of the pit type dwellings. The bedrooms of the cliff type, for example, average 7 by 3.2 meters, versus the pit type, which average 5.9 by 2.9 meters. The storage space average in the cliff type is 5.1 by 2.4 meters, while the pit type averages 4.1 by 2.2 meters. On the other hand, the kitchen of the pit type averages 5.4 by 2.5 meters, which is greater than the average of the cliff type, 4.5 by 2.3 meters. The cliff type dwellings do not consume space for entrances and stairways, while the pit type dwellings consume an average of 12.5 by 4.2 meters for such space. The total average of the cliff type rooms is greater (17.3 square meters) than the pit type average room (16.6 square meters). However the total average floor area of above-ground structures belonging to the cliff type dwellings was lower than that of the pit type dwellings (26.6 square meters versus 30.3 square meters).

Most of the cliff type above-ground structures are used for storage while most of the pit type above-ground structures are used as bedrooms. During our interviews the inhabitants mentioned that the

Table 3. Dimensions and Sizes of the Below- and Above-Ground Cave Dwelling Units of the Twelve Surveyed Dwellings.

Province	Dwelling Number	Dwelling Type	No. of Rooms (5)	No. of Residents (3)	Below Ground (net. w/o walls) (M = Meters)						Total Rooms M² (12)(6-11)
					Bedrooms M (6)	Storage M (7)	Kitchen M (8)	Other M (9)	Mix M (10)	Stairway/ Entrance M² (11)	
SHAANXI	1	Cliff	9	5	6.3x3.4 5.8x3.4	5.6x3.3 3.5x3.3 4.3x1.2 2.4x1.8	4.7x2.5	14.2x3.6 3.1x2.6			152
SHAANXI	2	Cliff	4	5	7.4x3.0 7.2x3.0 6.9x2.8		2.3x1.5				67
SHAANXI	3	Cliff & Earth-Sheltered	11	10	7.8x3.1 6.6x3.1 6.9x3.3 7.0x3.3 6.6x3.4	5.0x2.3 5.3x2.7 6.8x3.1			6.4x3.1 8.3x2.7 8.3x3.2		229
GANSU	4	Cliff	5	5	4.8x2.7	2.3x1.3 4.7x2.5	4.7x3.0	3.7x2.1			50
SHANXI	5	Cliff	7	4	5.6x2.4 5.7x2.0	6.3x2.7 5.6x2.1	6.3x2.3	3.3x2.8 1.7x1.7			80
SHANXI	6	Cliff	8	7	13.5x3.2 7.7x3.2	6.9x2.3 7.0x2.2 5.5x3.2 4.5x2.7 4.1x3.1 7.0x1.2			8.6x3.2		177
SHAANXI	7	Pit	11	11	3.0x3.0 5.8x2.7 5.1x2.7 6.3x2.7	3.5x1.8 3.7x2.3 7.0x2.8 4.6x2.7	6.7x2.7	6.7x2.7 2.4x1.4		3.0x1.5 4.5x3.2 18.9	161
GANSU	8	Pit	9	8	5.5x2.9 6.3x2.6 6.0x2.6	3.6x2.0 4.0x2.8		3.0x2.7	7.0x2.8 5.8x2.8 6.0x2.9	8.0x3.0 7.0x2.0 38	166
HENAN	9	Pit	11	11	6.3x3.0 6.2x3.2 6.2x3.1	1.2x1.0 8.6x3.2 2.0x1.4 3.1x1.8 4.5x3.0	3.1x2.1 8.0x3.0	2.0x1.6 1.0x1.0	3.3x2.0 6.0x1.8	17.4	161
HENAN	10	Pit	10	9	5.8x3.0 9.3x3.0 5.0x2.5 5.8x2.8 5.5x2.9	2.0x2.0 2.8x2.0 2.0x1.5			6.5x2.9	8.7x1.7 5.0x1.7 23.3	145
HENAN	11	Pit	9	8	6.9x2.5 7.6x3.1 6.9x2.5 6.0x2.8 5.8x2.6	7.7x2.7		9.5x2.5 5.8x3.5	6.6x2.6	7.5x2.3 17.2	189
HENAN	12	Pit	6	4	5.6x2.6 4.6x2.4	3.5x2.1 4.8x2.6 5.8x2.6	4.0x2.4			9.0x1.8 7.8x2.3 5.4x2.1 45.5	116
	Average, Total Average, Cliff Average, Pit		8.3 7.3 9.3	7.2 6.0 8.5	6.4x2.9 7.0x3.0 5.9x2.9	4.6x2.3 5.1x2.4 4.1x2.2	5.0x2.4 4.5x2.3 5.4x2.5	4.7x2.3 5.2x2.5 4.3x2.2	7.1x2.9 7.9x3.0 6.3x2.8	12.5x4.2 12.5x4.2	16.8 17.0 16.8

* Volume of earth excavated from rooms is calculated based on floor dimensions plus an average height of 2.5 meters from the floor to the highest point of the vaulted ceiling.

Table 3. Dimensions and Sizes, continued

Province	Dwelling Number	Dwelling Type	Below Ground, Con't		Cubic Meters			Above Ground (net. w/o walls)				
			Courtyard Floor M² (13)	Total Floor Area (14)(12+13)	Total Rooms M³ (15)	Courtyard Space M³ (16)	Total Space M³ (17)(15+16)	No. of Rooms (18)	Bed-Room (19)	Storage M (20)	Other M (21)	Total Floor M² (22)
SHAANXI	1	Cliff	172	324	397	731	1128	6	5.0x2.6	3.8x2.6 1.4x2.3 2.5x2.7 4.7x2.5	6.2x2.2	58
SHAANXI	2	Cliff	85	152	156	489	645					
SHAANXI	3	Cliff & Earth-Sheltered	412	641	231	90	321	6		2.6x2.6 1.2x1.0 3.3x1.7 3.2x2.7 2.3x1.5 2.9x2.1		31.3
GANSU	4	Cliff	378	428	99	1890	1989	3	5.2x3.3	3.3x2.4	3.5x2.7	34.5
SHANXI	5	Cliff	728	808	162	2868	3030	2		3.2x2.3 2.2x1.2		10
SHANXI	6	Cliff	509	686	403	2672	3075	3	4.3x2.6	5.3x1.3 2.7x1.3		25.6
SHAANXI	7	Pit	162	323	373	1216	1590					
GANSU	8	Pit	181	347	407	1172	1580					
HENAN	9	Pit	319	480	363	1847	2210	7	4.5x3.0 4.5x2.6 4.5x3.3 4.5x3.0 4.5x2.6		2.5x2.1 5.3x2.3	83
HENAN	10	Pit	58	203	266	⸗403	669	1			2.7x1.6	4.3
HENAN	11	Pit	136	325	369	1183	1552	1			3.8x1.4	5.3
HENAN	12	Pit	67	183	154	438	592					
Average, Total Average, Cliff Average, Pit			267.3 381 154	408.3 506.5 310.0	33 33 34	1051 1456 879.3	1531 1698 1365	2.4 3.3 1.5	4.6x2.8 4.8x2.8 4.5x2.9	2.8x2.1 2.8x2.1	4.0x2.0 4.4x2.1 3.6x1.9	31.6 32.0 30.9

above-ground bedrooms were primarily used in the summer and not in the winter. This seemed to be because of the high humidity and lack of ventilation in the pit type's below-ground rooms, while the cliff type's below-ground rooms enjoy slightly better ventilation and therefore less humidity.

The six surveyed dwellings of the cliff type (table 4) have an average of 2 families per complex, fewer than the pit type (2.6 families per complex). Similarly, the average number of residents per one complex in the cliff type is lower (6 persons per complex) than the pit type (8.5 persons per complex). The average floor area per resident was much higher in the cliff type (15.2 square meters) than in the pit type (4.3 square meters). The largest floor area per resident in the cliff type was 202 square meters, versus 44.3 square meters in the pit type, a condition that is unattainable for urban dwellers in China. The smallest space per person in the cliff type was 30 square meters, versus 22.5 square meters in the pit type. The total volume of soil dug out per resident in the cliff type was 55 cubic meters, versus 18.8 cubic meters in the pit type. The average floor size of the above-ground structure of the two types is reversed, with the pit type having an average of 30.3 square meters, versus 26.6 square meters for the cliff type. Finally, the average area per resident in the above-ground structure is 0.77 square meters in the cliff type, versus 0.18 square meters in the pit type.

The courtyards also differ between the two types (table 5), with the cliff type averaging 1 courtyard per complex, and the pit type averaging 1.5 courtyard per complex. However, the cliff type average is deeper than the pit type (10.5 meters versus 7.6 meters), with this differentiation being beneficial to thermal performance in the cliff type, which is also exposed to more sunshine and light. The cliff type courtyards are formed of soil dug from the cliff face and dwelling units.

The total volume of soil dug to form a cliff type courtyard is much greater than the pit type courtyard (1456 cubic meters versus 879.3 cubic meters). The average dimensions of cliff type courtyards are 23 by 19.8 meters, versus 13.3 by 8.7 meters for the pit type. The lengthiest courtyard is 33 meters in the cliff type, versus 20 meters in the pit type. The greatest single width of a cliff type courtyard was 26 meters, versus 18 meters for the pit type.

The previous findings and analyses described the vernacular Chinese house design principles; the cave dwelling distribution and constraints; their

TABLE 4. DENSITIES OF THE TWELVE SURVEYED DWELLINGS

DWELLING NUMBER	DWELLING TYPE	PERSONS AND SIZE							
				BELOW-GROUND				ABOVE-GROUND	
		NO. OF FAMILIES	NO. OF RESIDENTS	TOTAL FLOOR (meters^2)	FLOOR (meters^2) PER RESIDENT	TOTAL SPACE (meters^3)	VOLUME PER RESIDENT	FLOOR SIZE (meters^2)	AREA PER RESIDENT (meters^2)
		(23)	(3)	(14)	24 (14 ÷ 3)	(17)	(25)	(22)	(26)
1	Cliff	2	5	324	65	1,128	226	58	11.6
2	Cliff	3	5	152	30	645	129		
3	Cliff and Earth-Sheltered	3	10	650	67.8	321	32.1	31.8	3.18
4	Cliff	1	5	428	85.6	1,989	398	34.5	7
5	Cliff	1	4	808	202	3,030	757	10	2.5
6	Cliff	2	7	686	98	3,075	439	25.5	3.6
7	Pit	2	11	323	29	1,590	144.5		
8	Pit	4	8	347	43.2	1,580	197.5		
9	Pit	4	11	478	43.4	2,210	201	76	6.9
10	Pit	1	9	203	22.5	669	74.3	4.3*	0.48
11	Pit	4	8	326	40.7	1,552	194	5.3*	0.66
12	Pit	1	4	177	44.3	592	148		
Average, Total		2.3	6.6	408	10.7	1,531	41.5	28.4	0.47
Average, Cliff (Nos. 1–6)		2	6	508	15.2	1,698	55	26.6	0.77
Average, Pit (Nos. 7–12)		2.6	8.5	309	4.3	1,365	18.8	30.3	0.18

*Kitchen

Table 5. Courtyard Dimensions: Summary of the Twelve Surveyed Cave Dwellings

Dwelling Number	Dwelling Type	No. of Courtyards	Courtyard				
			Depth* (meters)	Length (meters)	Width (meters)	Floor (meters²)	Space (meters³)
			(27)	(28)	(29)	(13)(28 × 29)	(16)(27 × 28 × 29)
1	Cliff	1	8.5	16.5	9	172	731
				4	1.5		
				5	3.5		
2	Cliff	1	11.5	17	5	85	489
3	Cliff and Earth-Sheltered	1	17	33	25	412	90
4	Cliff	1	10+	21	18	378	1,890
5	Cliff	1	6	28	26	728	2,868
6	Cliff	1	10.5	23	18	509	2,672
				10	9.5		
7	Pit	2	7.5	(a)10.3	(a)9.3	(a)96	(a)718
				(b)9.5	(b)7	(b)66	(b)498
				19.8	16.3	162	1,216
8	Pit	1	6.5	15	11	181	1,172
				6.2	2.5		
				21.2	13.5		
9	Pit	3	9.5	(a)10	(a)7	(a)70	(a)666
				(b)14.3	(b)9	(b)129	(b)611
				(c)14	(c)7	(c)98	(c)465
				(d)4	(d)2.3	(d)9	(d)43
				(e)5	(e)2.6	(e)13	(e)62
				47.3	27.9	319	1,847
10	Pit	1	7	8	7.2	58	403
11	Pit	1	8.7	20	6.8	136	1,183
12	Pit	1	6.5	9.5	7.1	67	438
Average, Total		1.2	9	18.1	14.1	241	1,051
Average, Cliff (Nos. 1–6)		1	10.5	23	19.8	381 **	1,456 ***
Average, Pit (Nos. 7–12)		1.5	7.6	13.3	8.7	115.71	879.3

*Depth of Cliff patio represents the height of the cliff only.

**Patios of the cliff have different shapes (squares, triangles, etc.) which involved special calculation of each.

***Cliff space volume calculation is based on diagonal lines of the slopes and therefore differs from the pit calculation.

general thermal performance and basic design concepts; and, finally, overall specifications of the twelve cave dwellings researched. The twelve sites have been carefully selected to represent the different geographical regions of China's loess soil zone, and represent different design adaptations evolved under those local conditions. The following two chapters discuss the details in design pattern and thermal performance of each dwelling, with chapter 2 focusing on the six cliff type, and chapter 3 focusing on the six pit type.

2

CLIFF DWELLING CASES

This chapter is concerned with the case studies of the six cliff dwellings researched in the provinces of Shaanxi, Gansu, and Shanxi. It should be noted that a "region" in China is the second major administrative unit in the hierarchical structure, after province, and is followed by the county, township, and village. The township, also called the "commune," is divided into brigades that are further subdivided into teams.

Dwelling No. 1: Gao Ke Xi Family Home, Shaanxi Province

DESCRIPTION

The Gao Ke Xi dwelling in Gao Jia Team Village, Liquan County (fig. 15), 80 kilometers northwest of Xi'an, Shaanxi, was selected for study because of its orientation, its special combination of below- and above-ground structures, and because it is a prototype of the existing cave dwellings in this region. The dwelling is now owned by a farmer and was originally designed as a farmhouse. It is located near a main county road and near other cave complexes. The building site was selected primarily because of the nearby cliff and the availability of land. The cliff, however, is oriented toward the northeast, not the most desirable site because the dwelling receives less sunlight. The dwelling itself evolved over time; first the below-ground part was built and later the above-ground units. In perspective, the overall view is coherent in design and structure (fig. 16); however, the view from inside

the courtyard illustrates the combination of old and new styles. Above the cave dwelling there is an elementary school building. The location of the school minimizes rain penetration into the cave units. Access to the school, including a road and stairs, is above the livestock pens (fig. 16).

The form is typical of Chinese cave dwellings: only one end of each unit has a door, windows, and a view of the patio (fig. 17). Each cave unit used for residential purposes has two sections: one for living and sleeping (including a built-in heated bed) and the other innermost part, for storage (fig. 18). The front room is reasonably well lighted but the inner part receives much less. Some parts of the room are whitewashed, increasing light reflection into the storage room. The walls are cut through the loess soil and mortared with loess combined with straw. The floor is unpaved but has a hard earth crust. Window frames and doors are made of rough wood and walls and ceilings are usually lined with newspapers to minimize dust. In every sleeping room there is a bed platform constructed of blocks of earth with a covering of large bricks mortared with straw and mud. Several openings in the side enable a slow, straw-fed fire to be tended for a short time to heat the bed. A chimney leads outside.

The above-ground rooms of the Gao Ke Xi dwelling are located on the northeast end and consist of a salesroom to sell agricultural products, handmade shoes, etc.; a bedroom; storage space; and a reception area close to the entrance. Within the patio is a small structure used for storing wheat, rice, and other grain. It is elevated on bricks to prevent moisture from entering and a window is used for

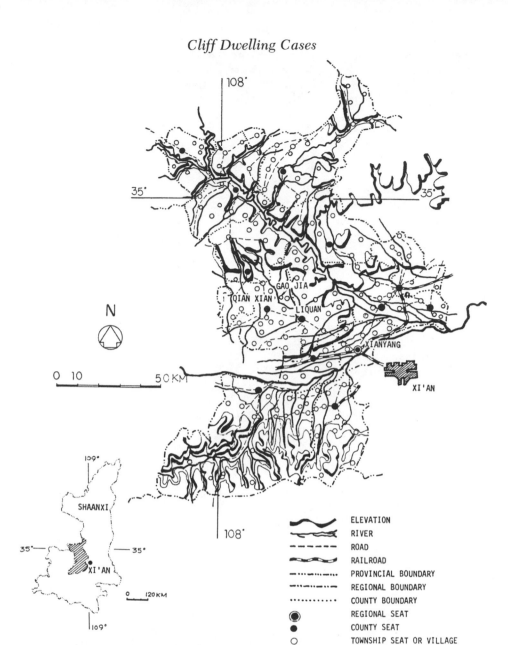

Fig. 15. Location of Gao Jia Team Village in Liquan County, Shaanxi province.

access since the structure does not have a door. The courtyard is deep on one side and level with the road on the other side (fig. 19). It is flat, unpaved, and landscaped with a few trees. The livestock space houses several pigs. The kitchen is located in the corner near the below-ground bedroom units and has two sections, one for storage and one for cooking and washing. It contains a coal-fired stove which is common in Chinese cave dwellings. The major problem with this, and with most other cave dwellings, is the lack of sunshine and ventilation. The dwelling itself is compact and meets the needs of the two families lodged there (fig. 20).

The climate in Liquan County is moderate: snowy and cold in winter, rainy and warm in summer. The snow accumulation rarely reaches one foot and the rainfall (annually about 600 millimeters) is heavy in summer and continues into the autumn. Drought is not unknown here and the region can be considered semi-arid. The prevailing winds are from the northeast and east, and the southeast and south. The summer rain comes from the east. The local inhabitants report that the climate in Shaanxi province seems to be changing, with widely fluctuating unseasonable temperatures in summer and in winter.

Fig. 16. Perspective view of the Gao Ke Xi family dwelling (No. 1), Gao Jia Team Village, Liquan County, Shaanxi province. The view is from the northeast side of the structure toward the cave-dwelling portion. Note the road and the stairway to the school located above.

Fig. 17. Courtyard of Gao Ke Xi family cliff cave dwelling (No. 1). Note the school atop the dwelling.

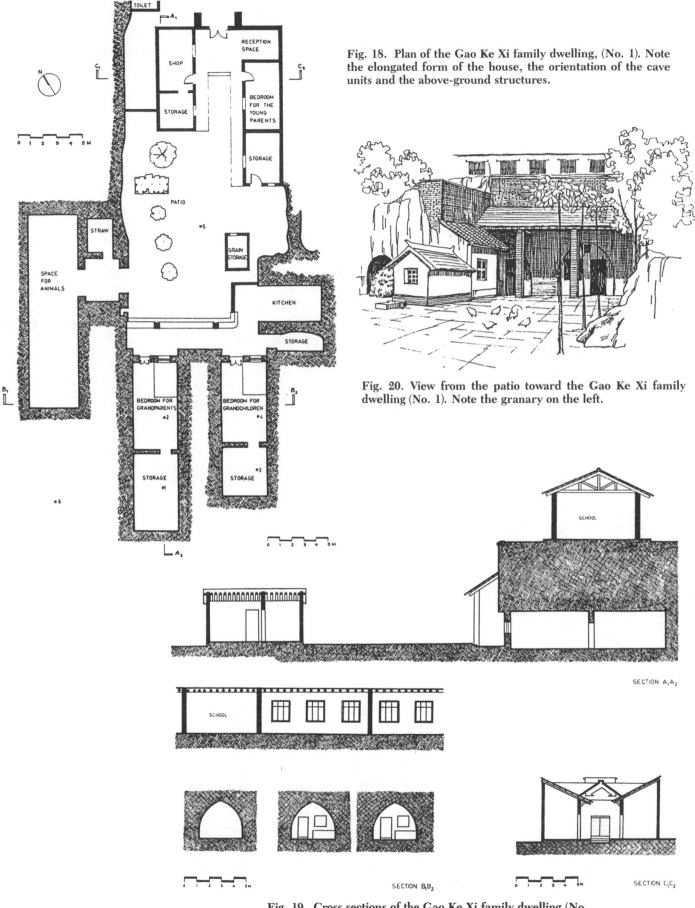

TOILET

SHOP

STORAGE

RECEPTION SPACE

BEDROOM FOR THE YOUNG PARENTS

STORAGE

N

PATIO

•5

STRAW

SPACE FOR ANIMALS

GRAIN STORAGE

KITCHEN

STORAGE

BEDROOM FOR GRANDPARENTS •2

BEDROOM FOR GRANDCHILDREN •4

STORAGE •1

STORAGE •3

•6

0 1 2 3 4 5 M

Fig. 18. Plan of the Gao Ke Xi family dwelling, (No. 1). Note the elongated form of the house, the orientation of the cave units and the above-ground structures.

Fig. 20. View from the patio toward the Gao Ke Xi family dwelling (No. 1). Note the granary on the left.

SCHOOL

SECTION A₁A₂

SCHOOL

SECTION B₁B₂

SECTION C₁C₂

0 1 2 3 4 5 M

Fig. 19. Cross sections of the Gao Ke Xi family dwelling (No. 1). The school building above the cave dwelling prevents rain penetration into the structure below. Note the differences in elevation among the cave dwelling, the patio, the portico, and the above-ground structure.

Topographically, the site is higher than its surroundings, offering a panoramic view of the farmlands from the dwelling entrance. The area is intensively cultivated by dry farming and irrigation farming, and drainage is good. There is no longer any natural vegetation or forest because of the intensity of the agriculture. There are, however, many tall trees that have been planted to mark avenues and roads or to divide the farmland. The soil is loess (yellow soil), which is rich and suitable for agriculture, but it is subject to erosion. The region is one of gently rolling hills and there is no acute differentiation in elevation.

There are twenty communes in Liquan County, each consisting of twenty-one or twenty-two teams. Each team contains a different number of families, the smallest consisting of 45 families with a total of 200 persons, and the largest, 800 families with about 3,200 persons. There are 489 teams in the county, and a total population of 350,000 people. Each team farms 316.2 acres (1,920 Chinese *mu*). Today 40,000 persons, or 11.4 percent, live in cave dwellings. The county is 54 kilometers long, and 31 kilometers wide.

There are cave dwellings in eight communes in Liquan County, all built into the cliffsides. In 1982, 667 new cave dwellings were constructed, making the total 21,542, comprising 495,460 square meters. The average cave dwelling room unit is 3.3 meters wide, 3.7 meters high and 6.6 meters long. The soil above the ceiling is 3.3 meters (1 Chinese *cha*) thick. Two-thirds of the cave dwellings are used as residences; others are used for manufacturing bricks, for housing animals, and for storing grain and fruit, mostly apples. Most of the dwellings face south to benefit from the sunlight. The loess soil encourages strong capillary action so many of the cliff walls are wet. The ceiling/roofs have to be at least 3 meters thick; otherwise collapse would be frequent. In some villages rain is the only source of water.

Gao Ke Xi and his wife, who live in Dwelling No. 1, like their cave dwelling because it is cool in summer and warm in winter. If they were to sell the complex (including the above-ground sections), they would ask 3,000 yuan for it (2.30 yuan equaled US$1 in 1984). In Gao Jia Team Village, fifty families (approximately 200 persons) live below ground. Older people of lower economic status are currently most likely to construct the less expensive subterranean dwelling units. Young families think it is old-fashioned to live that way, but they will use the caves temporarily during hot summers and cold winters. The young "rich" families prefer to build above ground to take advantage of the light. The lack of light is a major complaint of the below-ground inhabitants, with natural light being preferred over electric light.

THERMAL PERFORMANCE

To study the significance of heat exchange limitations in the cave dwellings, we measured the dry- and wet-bulb temperatures continuously for twenty-four hours and compared different parts of the units. We can extrapolate the humidity from these measurements as well.

At the Gao Ke Xi family dwelling, we selected six different locations for temperature measurements: four of them in underground rooms, the fifth in the courtyard and the sixth outside the building (figs. 21 and 22). The measurement began at 8:00 P.M. and continued every hour for the next twenty-four hours at each of the sites on 11–12 July 1984 for summer and on 9–10 December 1984 for winter temperatures.

Sites 1 and 2 are in a cave unit dug into the cliff a total of 12.7 meters and partitioned into two sections. The width of the rooms is 3.3 meters. The innermost room (site 1) is smaller than the front part and is used as a storage area. The front room is used as a living and bedroom with a heated bed, but without a heating stove. The ceiling is vaulted and is 3.1 meters high. Sites 3 and 4 are similar to Sites 1 and 2. The total length of Sites 3 and 4 is 10 meters. Site 3 is used for storage and Site 4 is a bedroom for the grandchildren.

The courtyard, Site 5, is quite wide and well lighted. It is a confined space and the center of all daily activities. It is unpaved and usually muddy in summer when rain is heavy. The temperature of the patio was measured in the center at a height of 0.75 meters. The patio contains a very few tall slender trees that do not interfere with ventilation, but neither do they provide much, if any, evapotranspiration. Site 6 was selected as being entirely outside the confines of the dwelling in the corridor of the school above ground. This site represents the general climatic conditions of the environs. The night and most of the day were cloudy when the summer temperature of Dwelling No. 1 was measured and it became rather drizzly in the afternoon. This condition moderated the temperature in the courtyard to some extent. (Note the wet-bulb temperature in the afternoon.)

The research measurements point out that there is a marked differentiation in temperature over the course of a twenty-four hour period with respect to

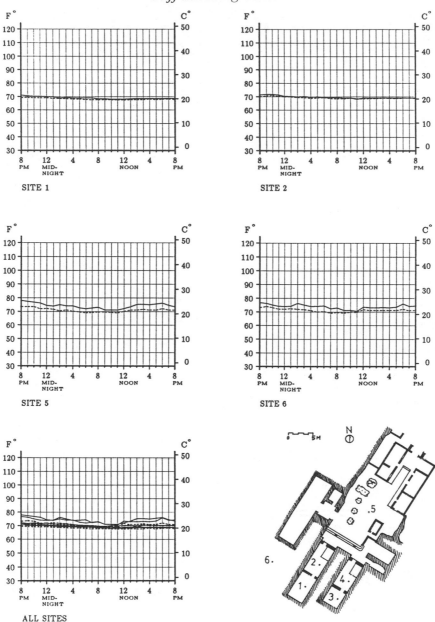

Fig. 21. Summer season dry-bulb (solid line) and wet-bulb (broken line) diurnal temperatures, 11–12 July 1984, in Gao Ke Xi family dwelling (No. 1). Temperatures were taken at six different indoor/outdoor sites. The findings shown are for four sites only: Site 1, the innermost room; Site 2, another below-ground room; Site 5, the patio; and Site 6, outdoor open space. Sites 3 and 4 are not shown since they are identical to Sites 1 and 2.

the subterranean rooms (Sites 1 to 4) relative to the courtyard (Site 5) and the outdoors (Site 6). Another important finding is that this distinction becomes acute on summer afternoons from 2:00 to 6:00 P.M. The latter is especially noticeable when we compare the courtyard with the cave dwelling itself. It is also noteworthy that the temperature is more comfortable at Site 6 than it is in the patio: the patio during those hours has the highest temperature of all six sites. This phenomenon repeated itself in almost all the other dwellings where temperatures were measured. Simply put, over the twenty-four-hour cycle, the Chinese cave dwelling courtyard is the worst place for activities at the peak diurnal

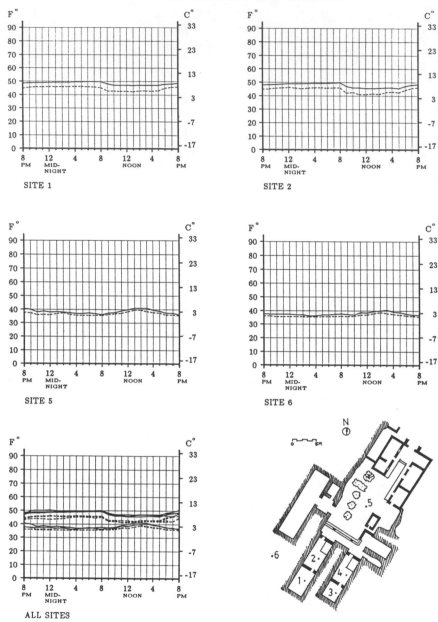

Fig. 22. Winter season dry-bulb (solid line) and wet-bulb (broken line) diurnal temperatures, 9–10 December 1984, in Gao Ke Xi family dwelling (No. 1). Temperatures were measured at six different indoor/outdoor sites, and four are shown here.

temperature and poses serious comfort problems that will be discussed in the concluding chapter.

The temperature process through soil is a complicated one. Although we can make some general statements about diurnal and seasonal heat gain and loss, it remains true that each site has its own unique character. This individuality results from different factors, such as the physical and chemical composition of the soil (sand vs. alluvial or rock, light vs. heavy, uniformity of composition, mineral content, etc.); percentage of humidity in the soil; level of the water table; cohesiveness of the soil; and, last but not least, the intensity of penetration of solar radiation.

The Gao Ke Xi family dwelling is 9 meters deep from the soil's surface to the floor of the dwelling and thus has almost stable diurnal temperatures throughout the summer season as well as the winter. On the other hand, there are noticeable differences among the dry-bulb temperatures of the four rooms compared to those of the patio (Site 5) and the outdoors (Site 6). Summer temperatures at the latter two sites fluctuate throughout the day, ranging between 23 and 26 C in the courtyard. More fluctuation occurs in the outside diurnal temperature (Site 6). Summer temperature variation among the four rooms is minor and all four sites range between 20 and 21 C. In sum, comparison of the dry-bulb temperature readings of all rooms and the patio indicate strong variations.

The pattern of the wet-bulb temperature readings is similar in general diurnal tendency. However, the wet-bulb temperature is always lower than

the dry-bulb. The differentiation between Sites 1 and 2, for example, is minor and follows the usual diurnal pattern. However, Site 1 (the inner room), has a lower temperature. A similar pattern is indicated in a comparison of Sites 3 and 4. Note that the range of all sites is between 20 and 22 C during the twenty-four-hour period.

In comparing the dry-bulb temperatures of Sites 1 and 2 with the outside temperature, we can clearly see the diurnal fluctuation of the outdoors contrasting with the relative stability of the inside readings. A similar trend is indicated in a comparison of the wet-bulb diurnal temperatures of all

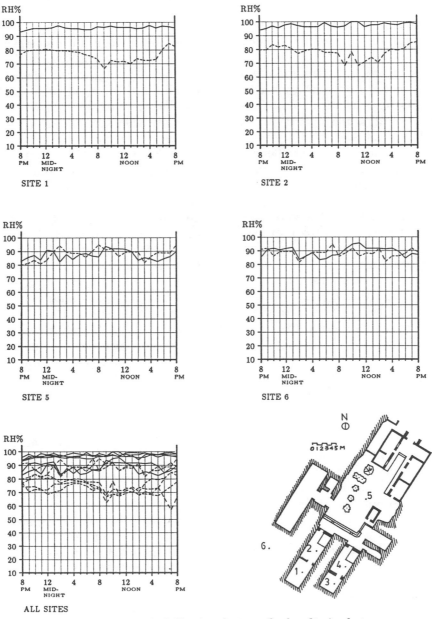

Fig. 23. Summer (solid line) and winter (broken line) relative humidity of Gao Ke Xi family dwelling (No. 1).

three sites with those of Site 6, outside.

A similar, but not identical, thermal performance pattern shows in the winter temperature measurement as well (fig. 22). The winter temperature of Site 1 ranges between 5 and 9 C for wet- and dry-bulb temperatures combined, compared with the summer temperature range of 20 to 21 C. However, it appears that the summer ambient environment of the below-ground units is more comfortable than the winter one. The rooms in the winter would require some extra heating, whereas in the summer air conditioning would not be required. It should be mentioned here that the doors and windows of Dwelling No. 1 are rough carpentry work and not hermetically sealed, thus there is an easy exchange of heat. Also, temperatures and fluctuations in Sites 1 and 2 are similar but Site 1 is a little warmer than Site 2.

As in the summer, there is a reaonsable winter temperature differentiation between Sites 1 and 2 compared with Site 5 (the courtyard) and Site 6. At Site 5, the lowest temperature occurs primarily during the second part of the night because of stagnated cold air. It begins to rise at about 8:00 A.M. and reaches a peak between 1:00 and 3:00 P.M. The temperatures of all sites (fig. 22) range between 1 and 10 C.

It should be remembered that the cave units of this dwelling face northeast, and thus they receive very little sunshine. It is more usual for the Chinese cave dwellers to select a southern orientation. To understand the wet-bulb temperature readings, one must compare them with the relative humidity of the air at the time of the measurement (fig. 23). On rainy days when relative humidity is high, human comfort is reduced. In this case study, most of the twenty-four-hour readings were made on cloudy days with precipitation beginning around 9:00 A.M. and continuing occasionally thereafter until 1:00 P.M. The difference in humidity between summer (high) and winter (low) within the below-ground rooms (Sites 1 and 2) is very noticeable (between 17–25 percent), due to the summer rain and condensation within the two sites. At Sites 5 and 6, there was much less difference in relative humidity between the summer and the winter.

Dwelling No. 2: Gou Shengzhi Family Home, Shaanxi Province

YAN'AN CITY CAVE DWELLINGS

Yan'an is located in Shaanxi, a province noted as

the cradle of Chinese civilization and the capital of thirteen dynasties (fig. 24). The region is mountainous, the soil loess, and the climate cold and snowy in winter and warm and rainy in summer (table 6). Yan'an City streches along the narrow valley of a rather shallow river (fig. 25). In the city, 40 percent of the people live in below-ground dwellings, and in the surrounding villages close to 90 percent do.

Most of the cave dwellings of the area are located on the steep slopes of the mountains and are either semi-above-ground terraced cliff cave dwellings, or earth-sheltered structures (fig. 26). Almost all the villages in this region are built on the slopes in the same style, even though the valley would have been a better building site (fig. 27). The intention was to save land for agriculture, and much of the land in valley and flood plain is cultivated. The few above-ground houses located along the river front are usually multi-level structures. The dwellings built further up the slopes have only one floor, either partially or entirely below ground.

Because of the availability of limestone resulting from the horizontal sedimentation, most of the buildings and many of the cave dwellings have incorporated it into their construction. While some ancient cave dwellings in the area seem to have been dug from stone, the prime rationale for building the cave dwellings seems to have been the workable loess soil itself, which has eroded along the cliff. The dwellers did not dig into the exposed stone but only into the loess, which ranges in thickness from 50 to 100 meters.

Yan'an City is a landmark of recent Chinese history since it is the city to which Mao Tse-tung led his Communist Revolutionary Army on the Long March—some 12,500 kilometers—arriving in October 1935. The city and its environs were the seat of the Party's Central Committee from 1935 to 1947.[9] Mao's first Yan'an City residence was a cave dwelling located near the river, not far from Dwelling No. 2 (fig. 28). On the northern side of the river there is another complex of cliff cave dwellings that was used by Mao's officers and by Mao himself. The latter lived in this complex from 1938 to 1943, and there he wrote fourteen books. An inner room was used as a shelter during Japanese bombing raids. At the present time, the semi-cave student dormitories of Yan'an University occupy a location on the northern slopes of the river.

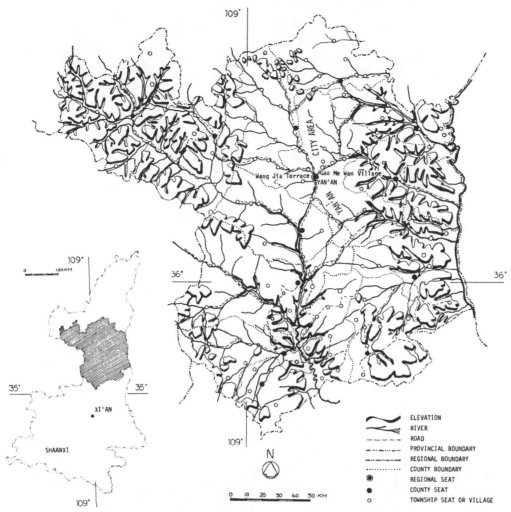

Fig. 24. Location of dwellings 2 and 3 in Yan'an City region.

Table 6. Temperature, Relative Humidity, Precipitation, and Wind in Yan'an City, Shaanxi Province.

| Month/Year | TEMPERATURE AVERAGE (C) | | | RELATIVE HUMIDITY (percent) | | PRECIPITATION millimeters | | WIND FREQUENCY (days per month) | | | WIND VELOCITY (km/hour) | | |
	Avg.	Highest	Lowest	Avg.	Lowest	Total	Highest in one day	Direction	Highest	Lowest	Avg.	Vel.	Dir.
Jan. 1982	−4.4	4.2	−10.3	45	7	0.3	0.3	SW or no wind / no wind	31	24	2. —	6.7	SW
July 1982	22.8	30.1	16.3	63	14	153.4	71.8	or SW	29	17	1.7	7.3	SW
Dec. 1982	−4.4	2.8	−9.5	48	20	0	0	SW		28	2.1	7.0	SW
Yearly Avg.	9.9	17.2	4.3	—	—	462.5	—	no wind or SW	29	19	—	—	—
Jan. 1983	−6.5	2.1	−12.6	45	0	0.4	0.2	WSW		30	2.2	5.7	ENE
July 1983	22.2	29.0	16.9	70	15	66.2	18.2	no wind or SW	25	18	1.6	8.3	NE
Dec. 1983	−4.4	2.8	−9.5	59	17	0.6	0.6	SW		31	1.9	5.3	NE
Yearly Avg.	—	—	—	—	680.7	—	no wind or SW	27	20	—	—	—	—

Notes:

In 1982: Rainy days were: 118 days; snow: 28 days; frost: 120 days. Snow stays on the ground: 16 days; hail: 2 days.

In 1983: Rainy days were: 126 days; snow; 15 days; frost: 139 days. Snow stays on the ground: 6 days; hail: 4 days.

Source: Yan'an Engineering Department.

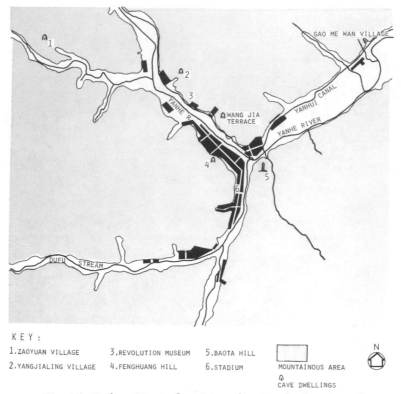

KEY:

1. ZAOYUAN VILLAGE   3. REVOLUTION MUSEUM   5. BAOTA HILL   ☐ MOUNTAINOUS AREA

2. YANGJIALING VILLAGE   4. FENGHUANG HILL   6. STADIUM   ⋒ CAVE DWELLINGS

Fig. 25. Yan'an City and vicinity, showing the location of Wang Jia Terrace neighborhood (Dwelling No. 2) and Gao Me Wan Village (Dwelling No. 3).

Fig. 26. Section of the southern slopes of Yan'an City, Shaanxi province. Note the combination of terraced cliff cave dwellings, semi-below-ground, and earth-sheltered dwellings.

Fig. 27. Cliff-dwelling neighborhood in the environs of Yan'an City near Yan'an pagoda.

Fig. 29. Perspective of the Gou Shengzhi family cave dwelling (No. 2), Wang Jia Terrace, west of Yan'an City, Shaanxi province.

Fig. 28. Mao Tse-tung's cave dwelling in Yan'an City, Shaanxi province.

Fig. 30. View of Wang Jia Terrace neighborhood. The Gou Shengzhi family complex is located in the center.

WANG JIA TERRACE AND DWELLING NO. 2

The Wang Jia Terrace neighborhood, where Dwelling No. 2 is situated, was built in the 1930s when Mao first came to Yan'an City. Before 1979, Dwelling No. 2 was used as a visitors' reception area, and since that time it has been used by families of people employed at a nearby school. The entire neighborhood is composed of cave dwellings (fig. 29). It is located on the northern slopes facing southwest, more than 100 meters above the river and about half a kilometer west of Yan'an City. Dwelling No. 2 contains three cave units occupied by members of the Gou Shengzhi family (fig. 30). The enclosed patio overlooks the river in front. On several surrounding terraces there are more cave-

dwelling units built in the same style.

The three cave units of Dwelling No. 2 are parallel to one another (fig. 31) and similar in dimensions, design, and style. The facades of the three units are covered with light-colored limestone and the front of each room unit is made of loess and bricks almost 1 meter high and covered with a thin layer of limestone and whitewashed. The patio is wide and accommodates chickens and storage space for wood and coal.

The eastern room, which is slightly longer than the others (7.5 meters by 3 meters) is used by the Gou Shengzhi couple. The height of the vault is 3 meters (fig. 32). The simple door is made of two pieces of rough wood. The floor is cut from loess soil and left unpaved. The window is made of thin,

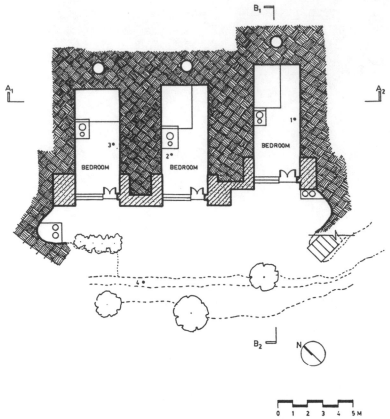

Fig. 31. Gou Shengzhi family dwelling (No. 2).

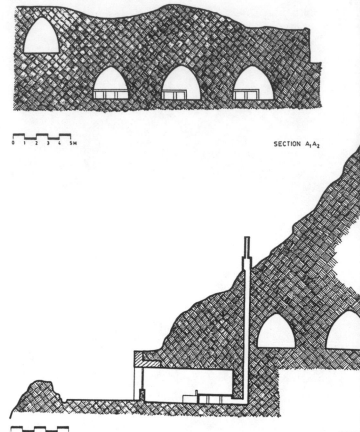

SECTION A₁A₂

Fig. 32. Southeast-northwest (A1, A2) and northeast-southwest (B1, B2) sections of Gou Shengzhi family dwelling (No. 2). Note the stove chimney at the rear.

latticed wood covered with a milky white paper which allows light to penetrate. Light is not a problem since the dwelling faces southwest. There is a small window in the uppermost part of the wall. The furniture is simple: a closet, desk, two tables, the traditional built-in heated bed, and a coal-fired stove on the left side. The arched ceiling, typical of Chinese cave dwellings, is supported by six rough-hewn log beams and crosspieces (fig. 33). The ceiling remains dry because of the thickness of the soil above. The middle and western rooms are occupied by relatives of Gou Shengzhi. These rooms are similar except that the stovepipe outlet of the middle room is located at the front of the unit, and the bed of the western room is situated crosswise at the rear. The stoves beneath the bed platforms are connected to chimneys constructed in the earth at the rear of the dwelling units. The privy is inconveniently located on a lower-level terrace and shared with many others.

Fig. 33. View inside one of the three room units of the Gou Shengzhi family dwelling (No. 2).

The monthly rent for one cave dwelling unit is 1.5 or 2 yuan (less than 1 dollar). Rent is usually calculated at 5 fen (1 cent) per square meter of living space per month. Heating fuel is scarce in the Yan'an region, so many inhabitants find it more economical to live in cave dwellings even though they have the option of moving to conventional above-ground houses. The empty cave dwellings in the neighborhood are reserved for students who live in them during the school year.

THERMAL PERFORMANCE

Since the three rooms of the Gou Shengzhi dwelling face southwest and are located high above the river, they have the benefit of light, sunshine, good ventilation, and air circulation. Four sites were researched both in summer and in winter, three indoors and the other, the outside courtyard. The three indoor room units are almost identical in size, orientation and positioning.

The summer dry- and wet-bulb temperatures of the three rooms (Sites 1, 2, and 3) are almost identical (fig. 34). Dry-bulb temperature is between 20 and 22 C, and wet-bulb temperature is between 18 and 20 C. The temperature of Site 2 shows less diurnal fluctuation than that of Sites 1 and 3. The patio (Site 4) has a lower temperature throughout the second part of the night than do the rooms because of cold air stagnation along the Yan He

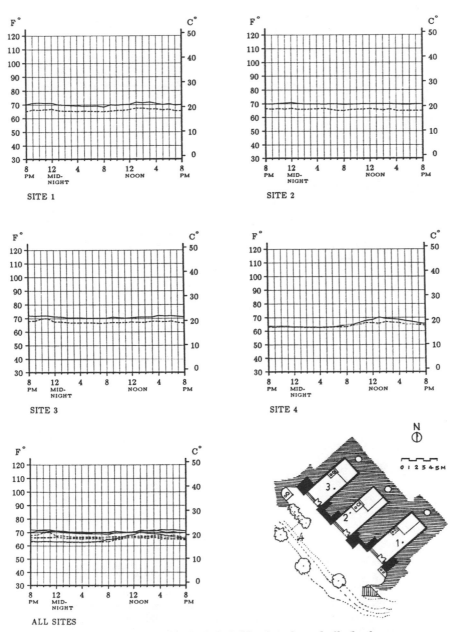

Fig. 34. Summer dry-bulb (solid line) and wet-bulb (broken line) diurnal temperatures (17–18 July 1984) of the Gou Shengzhi family dwelling (No. 2). Note the temperature stability of Sites 1, 2, and 3, and the fluctuation of Site 4.

River valley where Yan'an City is located. The temperature in the patio starts to rise in the morning and reaches its peak at 1:00 P.M. Note that dry- and wet-bulb temperatures are identical throughout the night because of the air stagnation, and are similar throughout the day, with an average differentiation of only two degrees, because of the rising air temperature and air movement throughout the valley. The temperatures of all sites show a fluctuation range of 5 degrees.

The winter temperatures of all four sites are quite in contrast to those of the summer (fig. 35). The dry- and wet-bulb temperatures of Sites 1, 2, and 3 are parallel and generally similar. The dry-bulb diurnal temperature fluctuates between 8 and 13 C with the lowest at the end of the night and the highest between 1:00 and 4:00 P.M. This condition requires a little heating within the rooms, especially during the second part of the night. This is a typical phenomenon for dwellings located in a

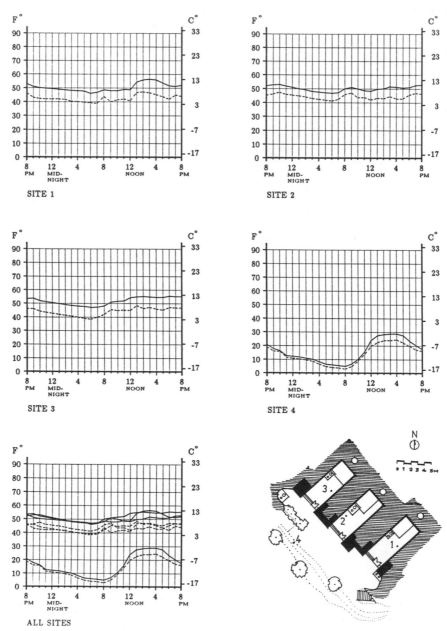

Fig. 35. Winter dry-bulb (solid line) and wet-bulb (broken line) diurnal temperatures, 8–9 January 1985, of the Gou Shengzhi family dwelling (No. 2). Note the temperature fluctuations of Sites 1, 2, and 3 compared with those of the summer. Greater differentiation is noticeable when comparing the patio (Site 4) seasonal temperatures.

valley such as this. The differentiation would have been greater if the house unit had been built above ground. The patio (Site 4) indicates great diurnal temperature fluctuation between night and day, and drops much below that of Sites 1, 2, and 3. During the night, temperatures drop sharply from −7 C in the evening to −17 C late at night. Both dry- and wet-bulb temperatures rise even more sharply in the early morning to reach a peak of −2 C between 1:00 and 4:00 A.M. before they again begin to drop.

In summary, the four graphs explain the superiority of the three inside rooms (Sites 1–3) at this cliff cave dwelling as compared to the conditions of the patio. We consider the outdoor diurnal temperature in the winter at this location to be climatically stressful. The graph of all sites (fig. 35) shows the great contrast between the patio (Site 4) and the three rooms. Note that although the dry- and wet-bulb temperatures are always parallel to each other diurnally, they are also isolated from each other by type.

The summer relative humidity is much higher than that of the winter (fig. 36). The condensation within the three cave dwellings and the rain in July and August cause this high increase. In the winter,

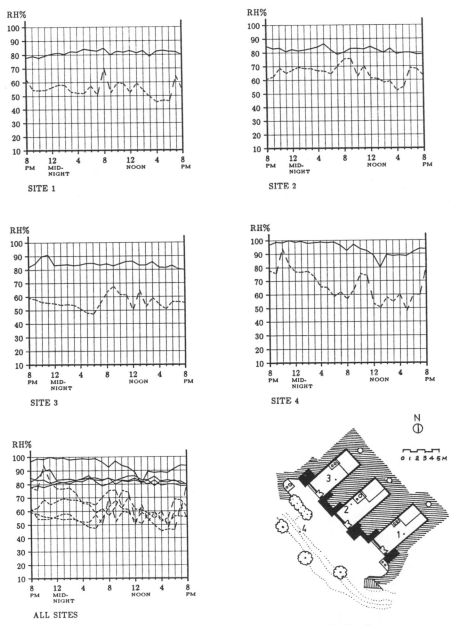

Fig. 36. Summer (solid line) and winter (broken line) relative humidity of the Gou Shengzhi family dwelling (No. 2).

relative humidity fluctuates very much, similar to the temperature fluctuations of the courtyard.

Dwelling No. 3: Zhang Yen Fu Family Home, Shaanxi Province

THE VILLAGE AND THE DWELLING

Gao Me Wan, where the Zhang Yen Fu family resides, is a village located on a slope along the river about 10 kilometers east of Yan'an City (fig. 25). The slope faces south and overlooks the green, fully cultivated valley. The population of the village is 400, comprising 80 families. The cliff cave dwellings are built on a series of semicircular terraces. The residents of Gao Me Wan village are farmers whose economic circumstances have improved with the agricultural reforms made by the government. Recently more dwellings have been constructed above ground, mostly stone houses and earth-sheltered habitats, an indication of modernization.

The focus of our research was the cliff cave dwelling of the Zhang Yen Fu family, designated Dwelling No. 3. It is built in an L-shape on a terrace above a major road (fig. 37). Four of the five east-wing units are completely below ground, whereas the west wing, also consisting of five units, is an earth-sheltered habitat (fig. 38). The east wing is covered by more than 15 meters of earth, with Sites 1 and 2—where temperature measurements were made—completely below ground. The west wing, at right angles to the east, is sheltered by only 2 meters of earth, and sunflowers are being raised on top. The west wing contains one below-ground storage unit located at the juncture of the two wings (fig. 39). Sites 3 and 4 are located at the end of this

Fig. 38. Part of the east wing of the Zhang Yen Fu family cliff cave dwelling (No. 3). Note the earth-sheltered roof section on the right side.

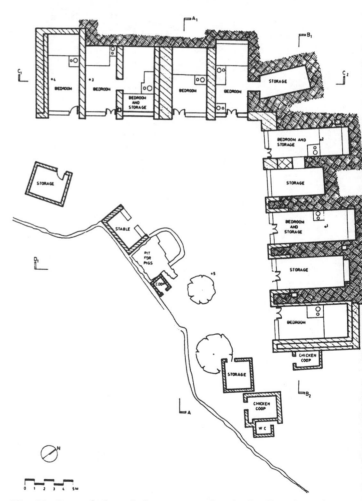

Fig. 39. Ground plan of Zhang Yen Fu family dwelling (No. 3).

Fig. 37. Perspective view of the Zhang Yen Fu family dwelling (No. 3) in Gao Me Wan Village, near Yan'an City, Shaanxi province.

wing away from the cliff. These designations enable us to compare the temperature measurements of the earth-sheltered wing with those of the below-ground wing (figs. 40 and 41).

The four indoor sites selected for measurement are almost identical in design; however, building materials vary. They each have the same kind of latticed window and rough wooden door at the front. The floors, of hard loess, are unpaved and the interiors are well lighted. The front walls of Sites 1 and 2 are made of stone while those of Sites 3 and 4 are of brick. The interior walls and arched ceilings of all sites are made of brick mortared with loess and straw and whitewashed.

The width of each of the cave units is around 3 meters, and eight of them have the typical built-in heated bed plus a stove for cooking (fig. 42). The

SECTION C,C₂

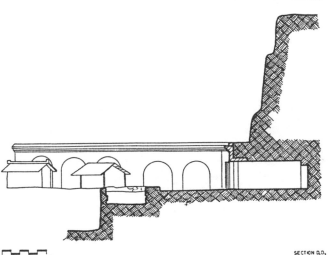

SECTION D,D₁

Fig. 41. Cross sections (C1, C2 and D1, D2) of Zhang Yen Fu family dwelling.

SECTION A,A₂

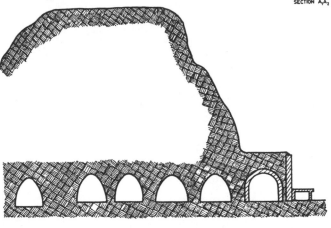

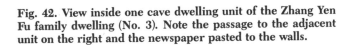

SECTION B,B₂

Fig. 40. Cross sections (A1, A2 and B1, B2) of Zhang Yen Fu family dwelling (No. 3).

Fig. 42. View inside one cave dwelling unit of the Zhang Yen Fu family dwelling (No. 3). Note the passage to the adjacent unit on the right and the newspaper pasted to the walls.

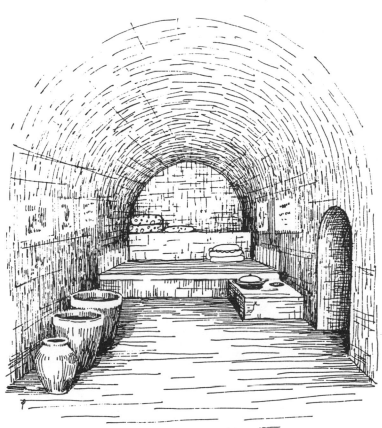

remaining three rooms are storage units. The two rooms in the east wing (Sites 1 and 2) were not being lived in at the time of our research and contained little furniture. Doors and windows were kept closed during the temperature measurements. The front walls of both sites were flooded with sunlight in the afternoon. The room connected to Site 2 was used to store grain and vegetables. Site 3 in the west wing was being lived in during the measurements. The screened window was often open, and the door was open most of the day as well, with a curtain covering the entrance. The connecting room was used to store food and the inner room's stove was used for cooking. Site 4 also was being lived in. The open doorway was covered with a curtain and the screened window was also open. The indoor stove was used for a short time. The room is well lighted during the daytime. Sites 3 and 4 received direct sunlight at noon but during the rest of the day was shaded.

The overall design of the complex is very efficient, with a large courtyard affording privacy and an attractive view overlooking the valley. The courtyard contains a few small animal pens and storage sheds that help to delineate the enclosure. The first four cave units of this dwelling were built in 1931 at what is now the intersection of the east and west wings. One room was added at the end of the east wing in 1967, and at the same time the owner faced the four older caves with stone. Later, in 1974, he built two more units, and in 1982 he added the rest. The owner did not cut openings in the back of the west-wing units because he does not like windows. All these expansions were made to keep together the growing extended family.

In the owner's estimate, to dig one cave dwelling unit costs about 100 yuan and takes about six days. As soon as the digging is finished, the window and door are added and he can move in. He pays 2–3 yuan per day for the digging, plus another 200 yuan for the bed, stove, door, and stone for the facade. The total cost of one cave dwelling unit comes to about 400 to 500 yuan, including marking of the cliff before digging starts. The soil removed from the site is used to grade and terrace the courtyard to keep rainwater from accumulating in front of the dwelling. Members of the family help with this work.

The owner said he enjoys building cave dwellings and advised the following:

1. The loess soil must be hard so the cave has stability. Hard soil requires more labor and is more expensive to dig, but soft soil should never be used for cave dwellings.

2. Check and select the proper orientation: never choose the north side because it is too cold and poorly lighted. The best orientation is toward the south.

Erosion is intense in this region and there are many gullies. The lower part of an eroded loess slope is usually stony which makes digging very difficult. Because of this, all the cave dwellings in Yan'an and its surroundings were constructed at a higher level—the mid-section of the cliff.

THERMAL PERFORMANCE

The Zhang Yen Fu dwelling is unique by virtue of being a combination below-ground cliff dwelling and earth-dwelling and earth-sheltered habitat. The right wing is oriented toward the southwest and the left wing toward the southeast, thus they receive plenty of light throughout the day. This type of house design provides a pleasing environment and does not produce a feeling of claustrophobia for the occupants, as is usually found among residents of pit cave dwellings.

Summer temperatures of the two subterranean rooms (Sites 1 and 2) are nearly identical and almost stable diurnally (fig. 43). The wet-bulb temperature graph is parallel to the dry-bulb during the twenty-four-hour period with a differentiation of 2.5 to 3 C. As is usual, the dry-bulb temperature is higher than that of the wet-bulb, yet both temperatures are within the comfort zone for this season and are much different from those of the courtyard (Site 5). The temperature at Sites 1 and 2 ranges from 17 to 21 C. Both temperatures are stable throughout the night and the early morning, and they increase slowly and steadily to a peak of 21 C (dry bulb) between 3:00 and 5:00 P.M.

In the earth-sheltered room (Site 3), temperatures fluctuate, especially during the daytime. The dry-bulb temperature reaches its peak between noon and 4:00 P.M. In comparing the two subterranean sites with Site 3, it becomes clear that the former provide a much better and more comfortable environment with more diurnal stability and lower temperatures.

The temperature of the patio (Site 5) shows much fluctuation between night and day. The dry-bulb temperature at night is similar but not identical to the wet-bulb, dropping from 19 C in the evening to 14 C at 6:00 A.M. On the other hand, the daytime dry-bulb temperature increases sharply toward the afternoon peak hours of 2:00 to 5:00 P.M., reaching 28 C. Thus, the dry-bulb temperature almost doubled between dawn and late afternoon, so typical of arid environments. This sharp difference between

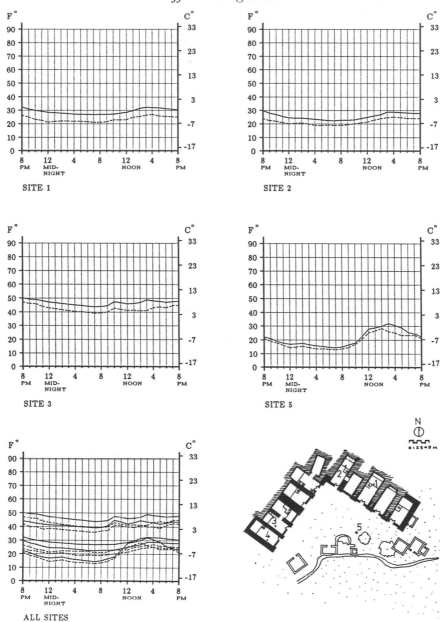

Fig. 43. Summer dry-bulb (solid line) and wet-bulb (broken line) diurnal temperatures, 19–20 July 1984 of Zhang Yen Fu family dwelling (No. 3). Note the differentiation in fluctuation between Sites 1 and 2 on the one hand (below-ground rooms) and Site 3 on the other hand (earth-sheltered).

the indoors and the patio contrasts even more acutely in the case of the pit cave dwellings. It is worth noting also that the temperatures of all sites range between 15 and 28 C.

Winter dry-bulb temperature, at a maximum of 9 degrees in Site 3, is still lower than the minimum of summer, 21 C (fig. 44). In any case, winter conditions here too will necessitate some heating during the night. There is a noticeable contrast between the earth-sheltered room (Site 3) and the two sub-

terranean rooms (Sites 1 and 2), which have lower temperatures. Both dry-bulb and wet-bulb temperatures of Sites 1 and 2 were diurnally below zero C while Site 3 was diurnally above 3 C. This is primarily due to the fact that Site 3 was heated.

Summer relative humidity is much higher than that of the winter in all sites (fig. 45). However, there is a smaller gap between the two at the earth-sheltered room (Site 3) than between below-ground Sites 1 and 2 due to the high winter relative humid-

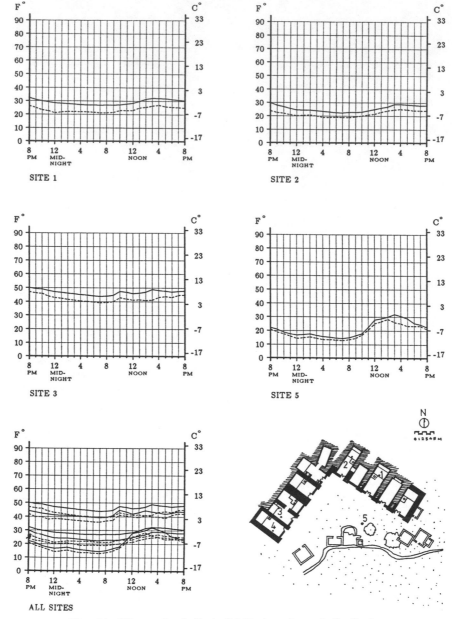

Fig. 44. Winter dry-bulb (solid line) and wet-bulb (broken line) diurnal temperatures, 10–11 January 1985, of Zhang Yen Fu family dwelling (No. 3).

ity of Site 3. We credit this to humidity absorbed by the outer walls during the summer and released indoors in the winter. The relative humidity of the courtyard (Site 5) was higher and fluctuated more than in any other site during both summer and winter.

Dwelling No. 4: Cao Yiren Family Home, Gansu Province

ENVIRONS

Gao Lan County, location of the Cao Yiren family dwelling, is 20 kilometers north of Lanzhou City, the capital of Gansu province (fig. 46). Lanzhou extends linearly for about 60 kilometers in a narrow valley bordering the Yellow River. There are many

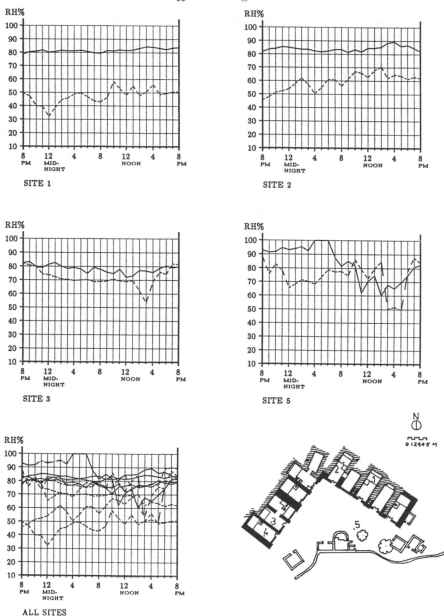

Fig. 45. Summer (solid line) and winter (broken line) relative
humidity of Zhang Yen Fu family cave dwelling (No. 3).

pagodas and pavilions in this area and some of the structures are placed on the slopes overlooking the river. The loess soil here is 100 meters thick. Vegetation has helped to control erosion, however, when the Cultural Revolution occurred, the planting of trees stopped. The city covers 146 square kilometers. South of the river, where most of the city is located, the land spreads out flat and wide. There is a new four-lane bridge over the river and, although controlled by a hydroelectric power dam 70 kilometers to the west where two or three branches converge, the waters are muddy with the yellow loess soil.

This is an arid region with a stressful climate. Rain is sparse, torrential, and of short duration. The main precipitation falls in the summer and the yearly total is less than 300 millimeters, not sufficient for dry-farming. Thus irrigation is required. This dryness is expressed strongly in the barren environment and causes the wind to be dusty and harmful to humans and animals alike. Life is difficult: at least one year of good weather is necessary to compensate for three years of drought. The hard dry loess soil amid the rolling hills supports the development of cliff cave dwellings that do not run much risk of collapse. Building materials used in

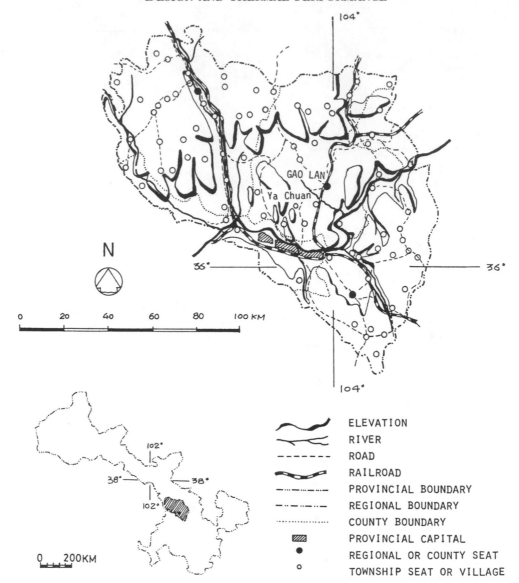

Fig. 46. Gao Lan County seat and Ya Chuan Village in Gansu province.

the region are primarily derived from the loess soil in the form of mud mixed with straw and sun-baked or burned bricks. Temperatures are high (28 C) from May to August and low in the winter (−18 C) from December to February. The diurnal temperature differentiation is large because of the aridity. This is a climate in which home design requires special techniques to combat the summer heat and the winter cold.

CAO YIREN DWELLING

The population of Gao Lan Village is primarily agrarian. The village's 254 people comprise 43 fam-ilies. Ten families live in cave dwellings combined with above-ground houses (fig. 47) and four or five families live only in below-ground dwellings. The unique quality of the Cao Yiren family dwelling (No. 4) is its location in an arid region, and it represents the farthest northwest dwelling re-searched. It is a cliff cave dwelling type where the four cave units all face southward (fig. 48).

The dwelling was constructed by the owner in 1978. It took him one month, working only in the evenings, and then he had to allow the units to dry out for two years before moving in. The major defi-ciencies in the cave dwelling, according to Cao Yiren, are the great difference between indoor and

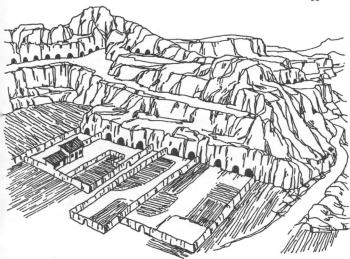

Fig. 47. The environs of Cao Yiren family dwelling (No. 4) in Ya Chuan Village, north of Lanzhou City, Gansu province. Note the cave dwellings terraced horizontally at the left and the barren environment surrounding the dwelling.

Fig. 48. Perspective view of Cao Yiren family dwelling (No. 4).

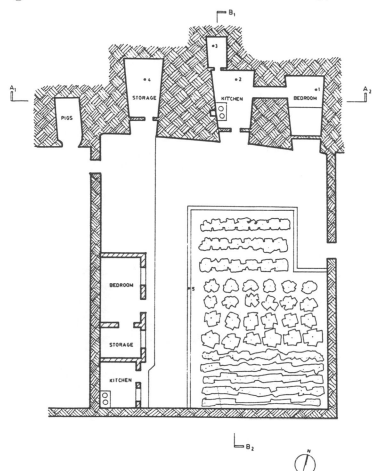

Fig. 49. Plan of Cao Yiren family house (No. 4).

outdoor temperatures, making it easy to catch cold or otherwise become ill; the low ceilings; and the small size of the rooms. The farmer-builder shares the residence with his wife and three children. He has close to one acre (5–6 *mu*) under cultivation. The cave units are arranged in a row facing the enclosed courtyard (fig. 49). The dwelling is sunny and light due to its orientation and the simplicity of its design. The ceilings of the rooms slant downward at the front (fig. 50), perhaps to minimize the effect of the strong outdoor albedo.

Four of the below-ground units of this family dwelling were investigated. Site 1 is the eastern room (fig. 51). It is entered by a passageway from Site 2, there being no door direct from the patio. The front wall, facing the patio, is entirely blocked by the traditional wide heated bed that occupies

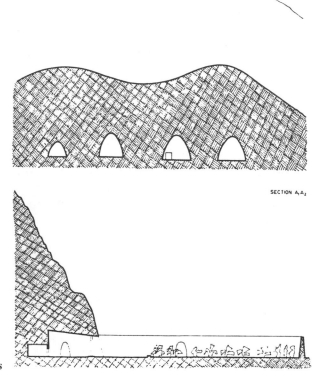

Fig. 50. East-west (A1, A2) and north-south (B1, B2) cross sections of Cao Yiren family dwelling (No. 4).

Fig. 51. The bedroom (Site 1) and the passage to the kitchen (Site 2) of Cao Yiren family dwelling (No. 4). Note the heated bed at the left occupying the width of the room.

Fig. 52. View inside kitchen unit (Site 2) of Cao Yiren family dwelling (No. 4). Note the passage to the bedroom at the right side and the entrance to the storage area (Site 3) at the back.

almost half the room. This small room also contains a low closet and a table. The window is large and was open much of the second half of the second day of the summer research period. The temperature measurements were taken deep on the east side of the room.

Site 2 is a kitchen with a closet and three stoves, two of which were in operation for most of the day (fig. 52). The site has a door to the courtyard, the aforementioned passageway leading to Site 1, and a curtained doorway to Site 3. The temperature measurements were made in the innermost part of the kitchen. Site 3, the inner room, is an extension of Site 2 and is a small, low-ceilinged storage area. Site 4 is located off the patio to the west and is also an area used to store grain and dry foods. The entrances to both Sites 3 and 4 are always curtained and the temperature measurements were made at the innermost points of each. All the below-ground sites are covered with loess soil to a depth of more than 15 meters.

Site 5 is in the middle of the patio that is enclosed on three sides by a wall 2 meters high and 0.40 to 0.60 meters wide made of loess soil. The patio, unroofed and sunny, contains a garden and, in the southwest corner, several newly-built above-ground rooms. Temperature measurements were taken in a shaded area near the garden. The main entrance to the dwelling complex is a low doorway on the east side of the patio. Another cave unit west of the patio enclosure is used as the pigsty.

THERMAL PERFORMANCE

Temperatures were measured at five sites of Dwelling No. 4, four indoors (Sites 1–4) and the fifth in the outdoor courtyard. Site 4 is not discussed independently because it is identical to Site 3 in thermal performance.

Summer temperature contrast between the outdoors and the indoors is acute because of the aridity of the climate, yet all indoor sites have diurnal

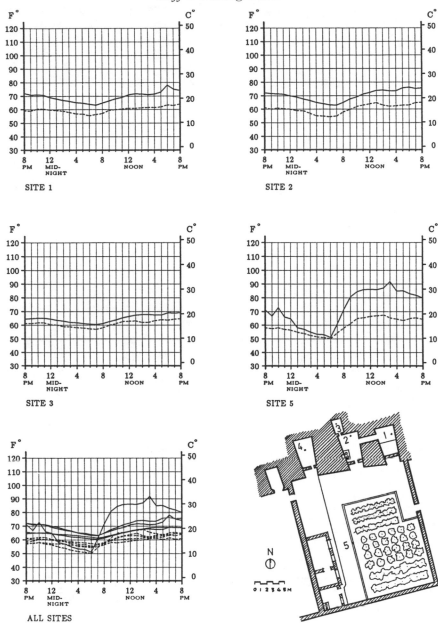

Fig. 53. Summer dry-bulb (solid line) and wet-bulb (broken line) diurnal temperatures, 26–27 July 1984, of the Cao Yiren family dwelling (No. 4).

temperature fluctuations (fig. 53), influenced by their southern orientation. Note that the front of the room unit is narrower than the back and seems to have been designed this way to minimize outside radiation penetration. In any case, set in a lowland surrounded by mountains, the courtyard of the cave dwelling generates much heat in the afternoons and consequently influences the temperatures within the four rooms. Both dry- and wet-bulb temperatures, which are diurnally parallel to one another, drop steadily and continuously from 22 C in

the evening to 17 C toward dawn and consequently bring cool nights. This is especially true late at night because of the cold stagnated air in the valley. Daytime temperature in Site 1, for example, rises quite sharply from 17 C in the morning to 26 C at 6:00 P.M. The kitchen (Site 2) has a similar diurnal temperature pattern especially at night. The temperature rise here during the day reaches a peak of 24 C between 5:00 and 6:00 P.M. The innermost room (Site 3), which was expected to have more stable diurnal temperatures, has a similar pattern;

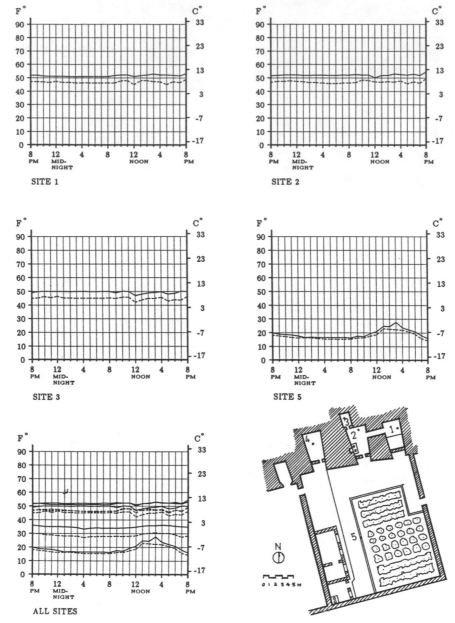

Fig. 54. Winter dry-bulb (solid line) and wet-bulb (broken line) diurnal temperatues, 14–15 December 1984, of the Cao Yiren family dwelling (No. 4).

yet both of its temperatures are lower diurnally and show less fluctuation than the other two rooms. This suggests that Sites 1 and 2 should have been dug deeper to achieve cooler temperatures on summer afternoons. The thermal performance of the patio (Site 5) is unique. It shows a sharp drop in both temperatures, from 22 C in the evening to 10 C by 6:00 A.M. Then it rises throughout the day to 33 C at 3:00 P.M., a difference of 23 degrees C between early morning and late afternoon. We attribute this

to the dry climate, the lack of vegetation cover in the patio and its surroundings, and the topographical formation of the valley. In any case, we can assume that the rooms would have been better off thermally if they had been oriented toward the southeast or the southwest. The temperatures of all sites, except the patio (Site 5) in the afternoon, range from 10 C at night to 25 C in the late afternoon, a differentiation of 15 degrees C diurnally.

Winter temperatures show much more stability,

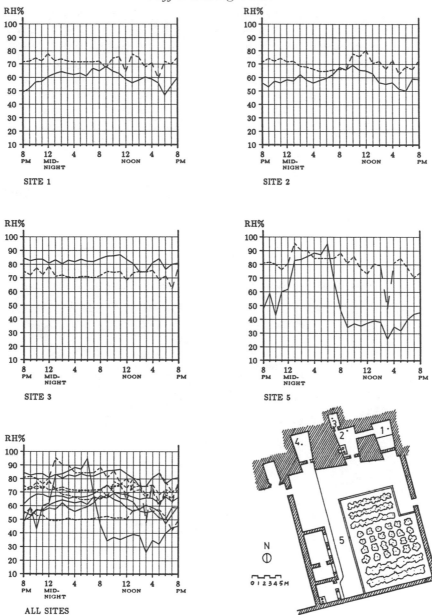

Fig. 55. Summer (solid line) and winter (broken line) relative humidity of Cao Yiren family dwelling (No. 4).

especially those of the indoor sites (fig. 54). The differentiation between dry- and wet-bulb temperatures is around 3 to 4 degrees C and the two temperature patterns are almost parallel to one another diurnally. The three rooms (Sites 1 to 3) are very stable throughout the night and are usually around 10 C (dry-bulb temperature), a condition that would necessitate some additional heating. The daytime temperature, on the other hand, has some minor fluctuation at noon and in the afternoon, yet it does not exceed 11 C. Still, a little heat would be required to make the rooms, comfortable. The patio

(Site 5) has a different diurnal pattern. There is a steady decline in temperature from the evening to the early hours of the morning when it is below freezing (−9 C). The daytime temperature increases sharply to its peak point of −3 degrees C at 3:00 P.M. both dry- and wet-bulb temperatures follow the same diurnal pattern and are very close to one another because of the low relative humidity in the air. The winter temperatures of all sites reach a maximum of 12 C and a minimum of −9 C. This great difference is attributed to the dryness of the climate.

The relative humidity of the dwelling complex, both in summer and in winter, differs from the other cave dwellings researched because of the extreme aridity and the very low precipitation during the summer. Evaporation is very high and puts the region in deficit of humidity. The relative humidity in summer is actually lower than that of the winter (fig. 55). This condition applies to all sites except the innermost room (Site 3). The increase in the sumemr relative humidity at the second part of the night in the two rooms (Sites 1 and 2) and in the courtyard (Site 5) is due to air contraction and the drop in temperature. The innermost room (Site 3) is less influenced by this temperature drop and its relative humidity stays stable during the night. Stable nighttime winter temperatures of the rooms keep the relative humdity stable as well. On the other hand, there is a sharp drop in the summer afternoon relative humidity in the two rooms (Sites 1 and 2) due to the heat increase. The winter afternoon relative humidity of those two sites fluctuates within a small range, similar to the winter condition of the courtyard in the afternoon. In the winter again, the innermost room (Site 3) is less influenced by outdoor changes and has the highest relative humidity. This is the only site where the relative humidity is higher in the summer than in the winter.

Dwelling No. 5: Zhao Qingyu Family Home, Shanxi Province

ENVIRONS

Taiyuan City, the capital of Shanxi province, is located in the north central part in a fertile valley 30

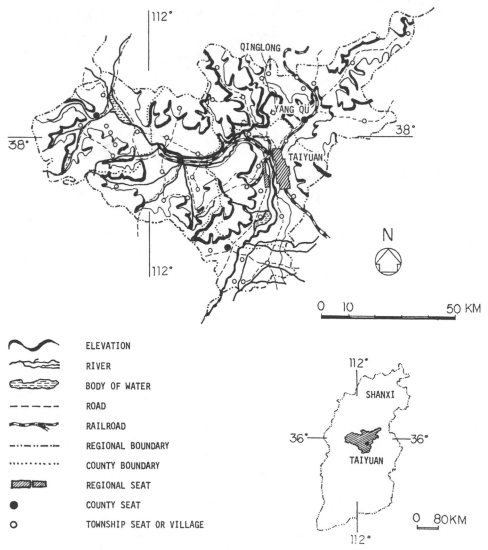

ELEVATION

RIVER

BODY OF WATER

ROAD

RAILROAD

REGIONAL BOUNDARY

COUNTY BOUNDARY

REGIONAL SEAT

COUNTY SEAT

TOWNSHIP SEAT OR VILLAGE

Fig. 56. Taiyuan City region where Yang Qu County is located.

kilometers wide and 120 kilometers long, north-south (fig. 56). This industrial region is one of the major coal centers of China.

The general climate of the area can be defined as warm and rainy in the summer and cold and almost dry in the winter (table 7). The rains begin here in early June and continue until September, with an average annual accumulation of 500 millimeters. Most of the winter precipitation is snow that falls in January and February. The end of July and the beginning of August are the hottest times, and usually January is the coldest month.

It is estimated that more than five million people live in cave dwelling villages in this province. Most dwellings are the cliff type, constructed in the mountains with stone fronts and terraced patios (fig. 57). The village of Tempo, for example, is around 10 kilometers northwest of Taiyuan City. The dirt road leading to the village is hard and dry. The loess soil is considered to be of medium strength. Here, some of the cave dwellings are constructed with two or three interconnected room units and only one entrance from the patio to the middle room, which is often the kitchen. Glazed jars are used for storage of food and there is a water tap in the village.

Another typical village is Zhen Zhu Mao in the mountains 30 kilometers northeast of Taiyuan City. For access, one climbs the mountain on a very rocky, dirt road. The village is built in terraced form

Table 7. Average Monthly Precipitation and Temperature in Taiyuan City Region (1983)

MONTH	PRECIPITATION (millimeters)	AVERAGE (C)
January	3.3	−6.8
February	5.6	−3.1
March	8.8	3.7
April	24.5	11.4
May	31.3	17.6
June	52.8	21.7
July	117.9	23.5
August	104.3	21.8
September	64.2	16.1
October	32.1	9.9
November	14.2	2.0
December	3.3	−5.0
Annual	462.3	11.8

Fig. 57. Cliff cave dwelling village and above-ground structures, Fushan County, Shanxi province.

overlooking the valley to the south with the cave dwellings oriented toward the south and southeast. In this village, there are 130 people, of whom 94, or 29 families, live below ground. The village is 300 years old, and it is only during the last two decades that the local people have started improving their cave dwellings. The villagers claim that the cave dwellings are wet in the summer, are not large enough for modern furniture and have a limited amount of air circulation. The cave units are well lighted inside, especially where they have been whitewashed. There is a door and a window at the front of each unit and the base below the window is faced with stone up to one meter high. The window occupies a large part of the front wall and is either of glass or thin white paper.

Almost all the non-cave dwellings in Zhen Zhu Mao have been built on the earth-sheltered system. These are attached to the cliff so that one or two sides are protected from heat gain and heat loss. After construction of the vault or arched ceiling, the roof is covered with a ten-centimeter layer of coal ash mixed with burned limestone (a product left over from the whitewashing) and with soil (in some places no soil is used). This cover mixture keeps the dwelling warm and prevents rain from entering. The earth-sheltered units have the same design as the cave dwellings. The facades are made of local gray stone since bricks are very expensive and must be brought from the valley or from Taiyuan City. The people have electricity, television, running water, and other conveniences in their homes.

ZHAO QINGYU DWELLING

The Zhao Qingyu family dwelling is located in

Qing Long Village, Yang Qu County, some 25 kilometers north of Taiyuan (fig. 56). The county population is 130,000, out of which 39,000, about 30 percent, live in cave dwellings. Qing Long Village numbers 1,000 people of whom 300 live below

Fig. 59. Bird's-eye view from the west of Zhao Qingyu family dwelling (No. 5), 25 kilometers north of Taiyuan City, Shanxi province.

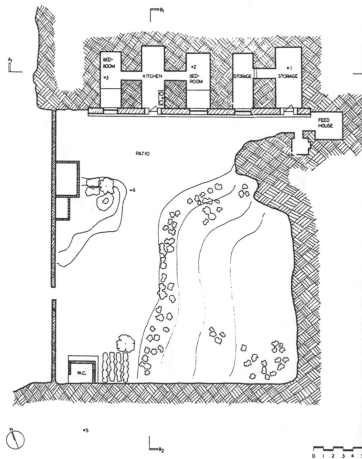

Fig. 60. Plan of Zhao Qingyu family dwelling (No. 5).

Fig. 58. The environs of Zhao Qingyu family dwelling (No. 5) in Qing Long Village, Yang Qu County, Shanxi province.

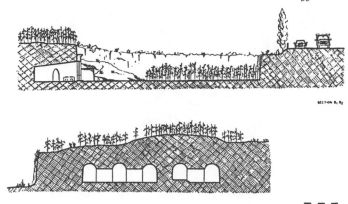

Fig. 61. Cross sections of Zhao Qingyu family dwelling (No. 5).

Fig. 62. View of the kitchen in Zhao Qingyu family dwelling (No. 5). Note the passages to adjacent bedrooms.

ground. The Zhao Qingyu family cliff cave dwelling is surrounded by many similar structures (fig. 58). The south side of the courtyard borders a major regional road; however, it is lower than the road itself. The five below-ground room units are placed along the northern cliff facing southward (fig. 59). A second cliff borders the patio on the east.

The design of this cliff cave dwelling is similar to that of the others in the area: a large patio, interconnected underground rooms, a limited number of entrances, and an isolated rest room (fig. 60). Three of the rooms are connected and have only one entrance from the patio through the central kitchen. The two adjoining bedrooms (Sites 2 and 3) feature the heated bed built the width of the room under the front window. The other pair of connected rooms have only one entrance and are used for storage.

The facade of the dwelling is covered more than three meters high with brick protecting the wall against landslides. The total height of the front part of the cliff is six meters (fig. 61). Inside the rooms, the ceiling slants downward a few degrees toward the rear to increase light penetration. The kitchen, one of the two longer rooms in the dwelling, is well lighted and used as the main entrance (fig. 62).

The patio is very large and is enclosed on the south and west by a wall more than two meters high made of loess soil. Much of the space in the patio is used for agricultural production and there is one entrance in the west wall.

THERMAL PERFORMANCE

Analysis of the summer thermal performance of the Zhao Qingyu dwelling can be divided into two parts: (1) the two below-ground rooms and (2) the courtyard and the outdoors (fig. 63). The two rooms (Sites 1 and 2) show a similar thermal performance diurnally. Site 1 is used as a storage area and has an almost stable temperature with the dry-bulb registering around 17 C and the wet-bulb about 15 C, each with a slight increase in the afternoon. Since the door of Site 1 is not frequently opened, the interior temperature stays relatively stable. The bedroom (Site 2), does not have an exterior door, and its situation adjacent to the kitchen may influence its thermal performance somewhat. The dry-bulb temperature is almost stable at around 21 to 22 C throughout the night and most of the morning. In the afternoon it is slightly higher by two degrees and drops again to 21 C in the evening. The wet-bulb temperature is rather similar, but three degree less, throughout the night and the morning. The afternoon wet-bulb temperature drops steadily until 4:00 P.M. and then drops sharply for one hour until it becomes stable toward night. The courtyard (Site 4) performs in a different pattern, almost identical to the outdoors (Site 5). The night temperature of the courtyard drops sharply from 22 to 15 C at 6:00 A.M. and then rises to a noontime peak of 29

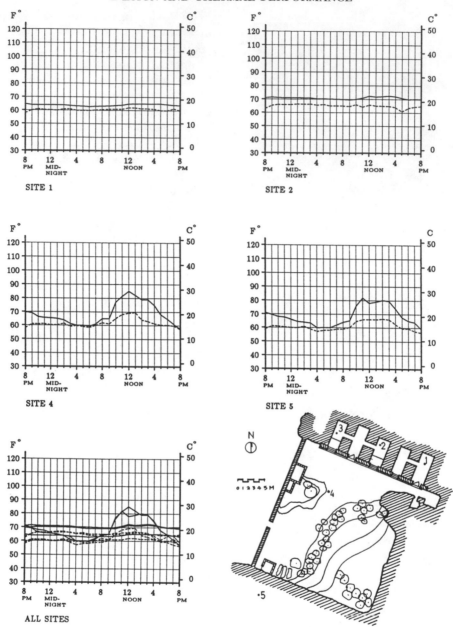

Fig. 63. Summer dry-bulb (solid line) and wet-bulb (broken line) temperatures of the Zhao Qingyu family dwelling (No. 5).

C. Thus, the differentiation between early morning and noon is 14 degrees C, which indicates tremendous change and instability. The temperature drops again quite sharply to 13 C at 8:00 P.M. These summer temperatures are typical of the pit cave dwelling patios but their appearance here in a cliff dwelling results primarily from the topographical forms enclosing the environs of the patio.

Winter temperatures present a pattern similar to that of the summer, especially in the patio (Site 4) and the outdoors (Site 5) (fig. 64). The two rooms (Sites 1 and 2) are noticeably different. Site 1, the storage room, has fluctuating temperatures of −12 (dry-bulb) and −15 (wet-bulb) at 8:00 A.M. and peak temperatures of −4 (dry-bulb) and −6 (wet-bulb) at 3:00 P.M. Both temperatures are below zero, the same as the patio (Site 4) and the outside (Site 5). The nearby cliff blocks off the sunshine. The other room (Site 2), is the only place where both dry- and wet-bulb temperatures are above freezing. We attribute this to heat from the kitchen stove and from the operation of the heated bed. In

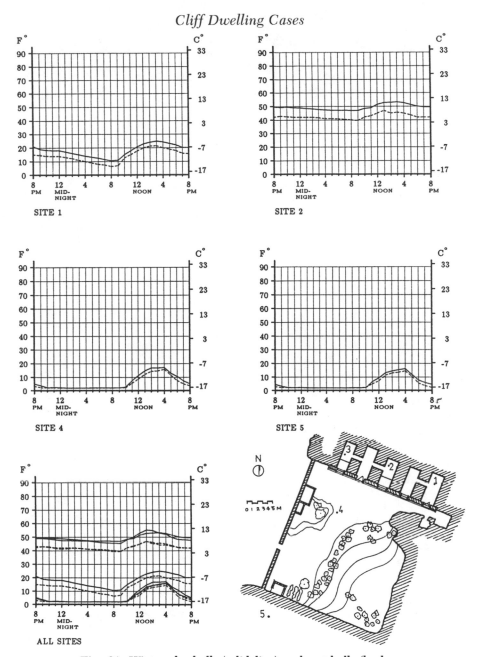

Fig. 64. Winter dry-bulb (solid line) and wet-bulb (broken line) temperatures of the Zhao Qingyu family dwelling (No. 5).

any case, the night temperatures drop from 10 C (dry-bulb) in the evening to 8 C at 9:00 A.M., and from 5 C (wet-bulb) to 4 C in the same period. The increase in both temperatures is steady, reaching 12 C (dry-bulb) and 8 C (wet-bulb) by the afternoon. The patio (Site 4) has identical stable dry- and wet-bulb temperatures throughout most of the night reaching a peak between 2:00 and 4:00 P.M. of −8 C (dry-bulb) and −9 C (wet-bulb); then they drop to −14 and −15 C respectively at 8:00 P.M. An almost identical development occurs at the outside location (Site 5).

Relative humidity is unique in the case of this dwelling, especially in the patio and the outdoors (Sites 4 and 5) (fig. 65). The two rooms (Sites 1 and 2) have higher relative humidity in the summer as compared to that of the winter, as can be expected. This is not the case in Sites 4 and 5 however, where the conditions are reversed. In Site 1, summer relative humidity falls between 80 and 90 percent during the twenty-four-hour period, while the winter relative humidity shows extreme fluctuation, from 28 percent at 6:00 A.M. to 78 percent at 2:00 P.M. Site 2, on the other hand, has a relative humid-

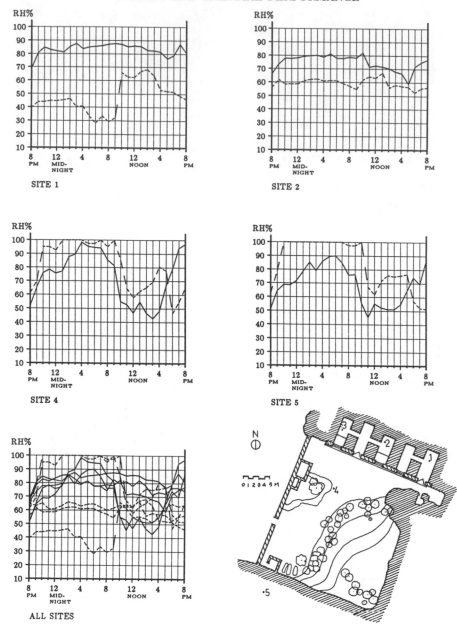

Fig. 65. Summer (solid line) and winter (broken line) relative humidity of the Zhao Qingyu family dwelling (No. 5).

ity in the summertime of around 80 percent during most of the night and part of the early morning and drops down to 58 percent at 5:00 P.M. Winter relative humidity fluctuates less, measuring around 60 percent during the night. It rises to 67 percent at 1:00 P.M. and drops to 53 percent at 6:00 P.M. The patio's summer relative humidity is in extreme fluctuation, rising sharply during the night to 98 percent at 4:00 A.M. and then dropping precipitiously to 43 percent at 3:00 P.M. The winter relative humidity is higher than that of the summer. It reaches 100 percent during the second part of the night and

then drops to 62 percent at noon and to 46 percent at 6:00 P.M. A pattern similar to that of the patio developed in Site 5.

Dwelling No. 6: Cui Mingxing Family Home, Shanxi Province

ENVIRONS AND DESIGN

The Linfen region is located among the rolling hills and valleys of southern Shanxi province.

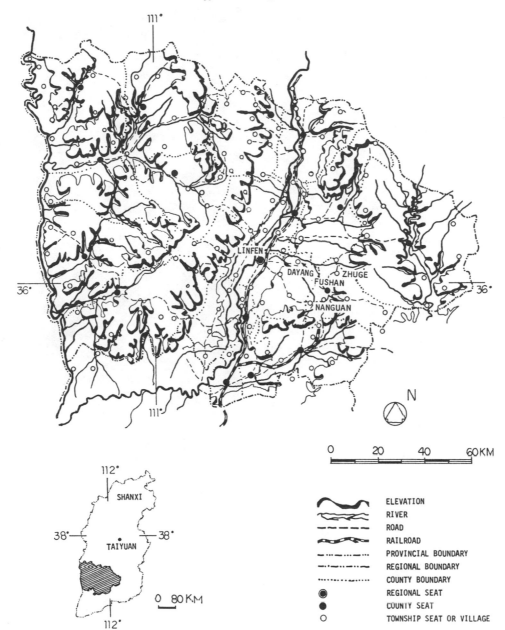

Fig. 66. The location of Dayang Village in Linfen region, Shanxi province.

Linfen City is set in the lower part of a valley near the Fen He River, a tributary of the Huang He or Yellow River (fig. 66).

The climate is warm and rainy in the summer, cold and snowy in the winter. Annual precipitation ranges from 400 to 600 millimeters. Although Shanxi is one of China's most industrialized provinces, the Linfen region is primarily devoted to agriculture and is not advanced technologically. The loess soil ranges between medium and hard, suitable for construction of the large concentration of cave dwellings in the area.

In Yang Shue Village, 25 kilometers north of Linfen City, for example, a broad range of the population lives in caves. The village is located on a sloping site in South Valley (the same valley in which Linfen City is located). The loess soil is harder than that of Taiyuan, and the percentage of limestone is high. According to the local Architectural Society, 300 to 400 of the 1,600 people in Yang Shue dwell in below-ground habitats. Many of the cave dwellings are more than 300 years old; one

Fig. 67. Earth-sheltered experimental dwelling constructed by the Architectural Society of China near Linfen City, Shanxi province. Note the corn growing on top of the building.

Fig. 68. View to the courtyard from the bedroom, dwelling in Fushan County, east of Linfen City, Shanxi province. Note the heated bed structure.

ancient dwelling unit measured 8 by 3 by 3.5 meters, essentially the same dimensions often used today. Improvements in the general economy have helped to increase construction, and many new cave dwellings have been built in the last decade. Before the Communist Revolution, the image of below-ground living was decidedly negative. A common saying was that, "Rich people live in houses [above ground] and the poor live in caves." Today, however, this view has changed somewhat and experiments and research are being conducted to upgrade cave design and living conditions (fig. 67).

Another example is Fushan County, 40 kilometers east of Linfen City, where cave dwellings predominate. The 120,000 people of the county are involved in the production of iron ore, coal, wine, and silk. According to local officials, 80,000 residents, or 70 percent of the population, live in cave dwellings. Most of the dwellings are built into the sloping sides of gullies and ravines. The improvement in the economy and the general increase in living standards has accelerated the rate of construction of additional cave units with about 400 to 500 dwelling complexes being added annually (fig. 68).

Northeast of the county seat of Fushan, Zhuge Village contains a cave dwelling with a long history. One of its innermost rooms contains a food storage well amost 1 by 1 by 2 meters deep, surmounted by a brick arch. Another cave unit in the village measures 5 meters wide, 12 to 15 meters long, and 4.5

Fig. 69. Four levels of cliff cave dwellings in Nanguane Village, Fuhsan County, 40 kilometers east of Linfen City, Shanxi province.

meters high, exceptional dimensions. The soil is 6 meters thick above the dwelling and there is a small second-story cave above the lower units. There is much light in the dwelling because of the height of the ceilings and the large number of windows.

Nanguan Village, west of the town of Fushan, is built into the cliffs on either side of a gully, one side facing north and the other south. The total population of the village lives in cave dwellings. The Architectural Society of Shanxi is conducting research at this location, measuring both the wet- and the dry-bulb temperatures in one of the cave dwelling units which faces north. The room is 9 meters long and 4 meters high and the walls are moist. In Nanguan, other cave dwellings have been constructed four stories high (for different families) along the sides of the ravine (fig. 69), consequently some face south and enjoy sunlight and less humidity on the walls and ceilings. Others face northward and have serious problems of lack of light and too much moisture. Another cave dwelling in the village has a room unit located 4 meters below the highway. This room is 8 meters long by 3 meters wide and high, with 5 meters of soil above. The roadway is hermetically sealed with asphalt that prevents water penetration into the soil. However, there can be some problems resulting from vibration from vehicle movement.

The traditional Chinese semi-enclosed house is expressed in another cave dwelling where three room units are built as three sides of a hexagon, with the central unit facing southwest, and the other two facing southeast and west. Thus they benefit from a maximum amount of light and sunshine during the day.

The dwelling of the Cui Mingxing family is located in Dayang Village, some 20 kilometers northeast of Linfen City. The village is part of Dayang Commune, which contains 17,000 people in 22 villages and 19 brigades. In the entire commune, there are 2,000 persons living below ground. Dayang Village is prosperous, with electricity and a paved asphalt road. Out of 3,000 people, one-sixth of the population and mostly older people, live in cave dwellings. No new cave dwellings have been constructed since 1971. The local explanation is that the cave dwellings do not respond well to changes in the standard of living and that they are not lighted well enough for modern needs.

The dwelling that was researched belongs to Cui Mingxing. The owner is the secretary of the Architectural Society of the Linfen region and he himself now lives in Linfen City, while three of the older members of the family live in the dwelling in Dayang Village.

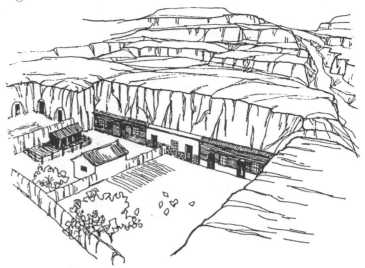

Fig. 70. View of Cui Mingxing family dwelling (No. 6) and its environs in Dayang Village, 20 kilometers northeast of Linfen City, Shanxi province.

Fig. 71. Bird's-eye view from the south of Cui Mingxing family dwelling (No. 6).

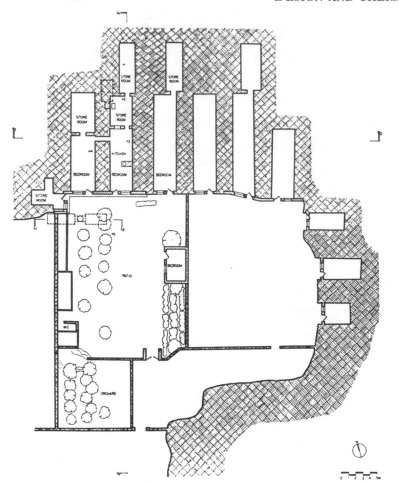

Fig. 72. Plan of Cui Mingxing family dwelling (on the west) and the neighboring dwelling (on the east).

Fig. 73. View from the westernmost bedroom (Site 4) toward the inner storage room, Cui Mingxing family dwelling (No. 6). Note the underground passage to the kitchen on the east.

The Cui Mingxing dwelling was constructed about seventy years ago in a terraced cliff environment (fig. 70). There is an earthen wall separating this dwelling complex from its neighbor to the east. Overall, the cave units are well integrated with the cliff and the adjacent flat area (fig. 71). The three main room units are on one side of the cliff facing south toward a huge patio measuring 18 by 23 meters. Most of the room units have additional inner storage areas. The cave dwelling adjacent on the east is similar in this respect (fig. 72).

Two of the three main dwelling units are very long, with the 20 meter length of the central unit divided into three sections. The innermost section (Site 1) is short of light and ventilation and is used for storage only. The middle section (Site 2) is used for storage also and has a below-floor-level space or well in the rear that may have been used in the past for concealment or for storage. The vault of the

main kitchen/bedroom unit (Site 3) is supported by beams and mortared. The west bedroom (Site 4) is well mortared and paved with bricks; the vault is covered with gray bricks (fig. 73). The inner extension is used for storage.

The patio (Site 5) is very large and, along with the adjacent dwelling complex on the east, is surrounded by the cliff on three sides (fig. 74). It is made up of a large unit (20 by 20 meters) and a smaller section (10 by 9.5 meters). The overall depth is 9 meters (fig. 75). On the east side of the main patio, there is an above-ground structure used as a bedroom and on the west side, the privy. There is a food storage well ten meters deep located at the northwest corner. It is used to store sweet potatoes and other vegetables and fruit. The cool temperature within the well keeps the food from decay. An orchard occupies the smaller patio enclosure. At the main entrance to the large patio there is an

Fig. 74. View from the south: patio of Cui Mingxing family dwelling (No. 6).

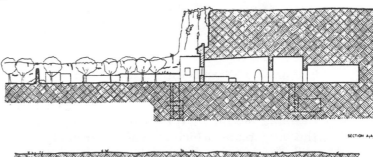

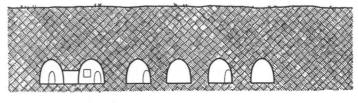

SECTION A₁A₂

SECTION B₁B₂

0 1 2 3 4 5M

Fig. 75. Cross sections of Cui Mingxing family dwelling (No. 6). Note the food storage pit located in the northwest corner of the patio and the well located inside Site 2.

0 1 2 3 4 5   SECTION C₁C₂

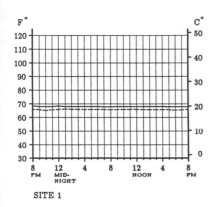

SITE 1

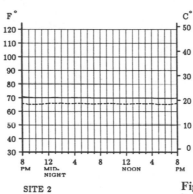

SITE 2

Fig. 76. Summer dry-bulb (solid line) and wet-bulb (broken line) temperatures of Cui Mingxing family dwelling (No. 6). Note the neighboring cave dwelling complex on the east.

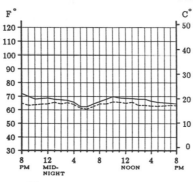

SITE 3

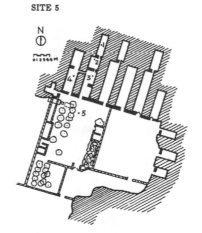

SITE 5

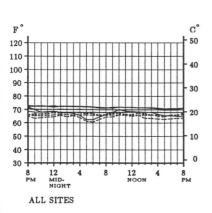

ALL SITES

additional transitional courtyard, a design element in keeping with the traditional Chinese house concept.

THERMAL PERFORMANCE

The significance of the Cui Mingxing family cave dwelling lies in its twenty-meter long rooms and the large patio, which is nearly surrounded by the cliff, leaving the open part for air circulation and heat exchange.

Summer temperatures of the three sections of the cave unit measured (Sites 1, 2, and 3) are similar to one another but not identical (fig. 76). Site 4, the bedroom in the northwest corner, has the same temperatures as Sites 2 and 3 and therefore has not

been included in the graphs. The coolest and most stable location is Site 1 because of its deep penetration into the cliff. The temperature fluctuates and increases as one moves from Site 1 to Site 2 to Site 3. The dry-bulb temperature of Site 1, for example, is 20 C diurnally; Site 2 is 21 C and Site 3 is between 21 and 22 C. A similar pattern is shown by the wet-bulb temperature. The patio (Site 5) fluctuates very much. The dry-bulb temperature drops throughout the night to 17 C between 5:00 and 6:00 A.M. and then begins to rise to a peak of 21 C at 10:00 A.M. It drops to 18 C toward evening. Wet-bulb temperatures, while lower, show a similar pattern.

Winter temperature patterns of the indoor sites are similar to those of summer (fig. 77). Site 1 has a

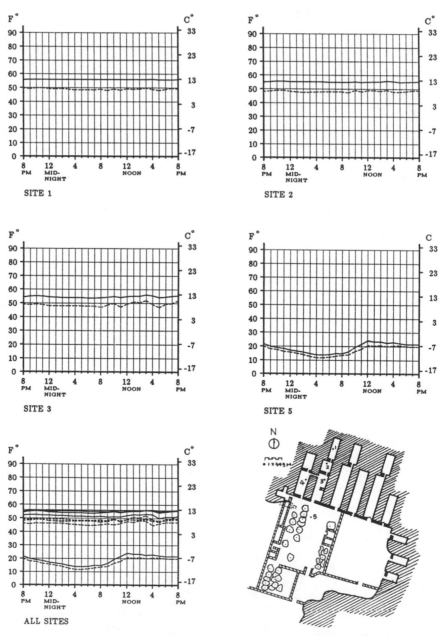

Fig. 77. Winter dry-bulb (solid line) and wet-bulb (broken line) temperatures of Cui Mingxing family dwelling (No. 6).

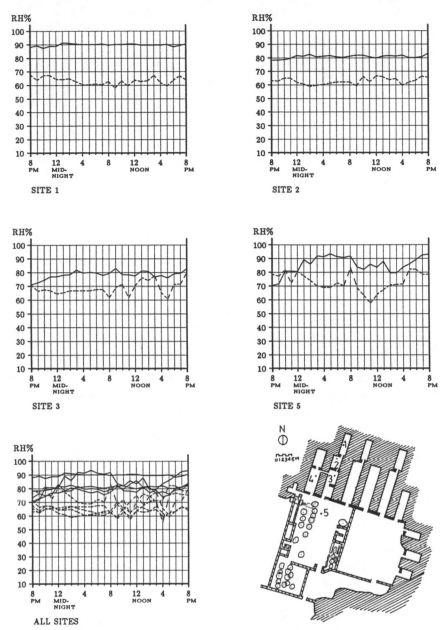

Fig. 78. Summer (solid line) and winter (broken line) relative humidity of Cui Mingxing family dwelling (No. 6).

stable diurnal dry-bulb temperature of 13 C and wet-bulb temperature of 9 C. Site 2 has a lower temperature of both dry-bulb (12.5 C) and wet-bulb (8.5 C). Site 3, on the other hand, because of its closeness to the patio, has developed diurnal fluctuations in both temperatures, which register lower than those of Sites 1 and 2 (around 11 to 12 C). The patio (Site 5) is far lower than the other three room sites showing the lowest temperature of −10 C late at night and its peak of −6 C at noon. The temperatures of all sites except the patio range from 7 to 13 C, showing explicitly the great contrast in comfort between the indoor temperatures and those of the patio in the winter.

The summer relative humidity of all indoor sites is higher than that of the winter (fig. 78). The rela-

tive humidity of the innermost room (Site 1) is around 90 percent, that of Site 2 is above 80 percent most of the time, and that of Site 3 fluctuates between 70 and 80 percent. The winter relative humidity fluctuates more than the summer's, especially during the daytime when Sites 1 and 2 both register between 60 and 70 percent, and Site 3 is between 60 and 78 percent. The patio, as usual, introduces a different pattern. In the late evening, the summer relative humidity is lower than that of the winter, yet all of the rest of the time, the summer relative humidity is higher than the winter: between 80 and 94 percent. The winter relative humidity fluctuates between 58 and 83 percent.

3

PIT DWELLING CASES

The chapter is concered with the pit cave dwellings researched. These are six case studies covering dwellings located in different regions of the provinces of Shaanxi, Gansu, and Henan.

Dwelling No. 7: Bai Lesheng Family Home, Shaanxi Province

ENVIRONS

Qian Xian County, where the research on Dwelling No. 7 was conducted, is located 75 kilometers northwest of Xi'an City, Shaanxi. Of the twenty-two communes there, nine have cave dwellings. Almost half of the county is hilly and mountainous, ranging from 600 to 1,400 meters above sea level. The other section is a valley with an elevation of 400 to 600 meters. Most of the cave dwellings are located in the mountains, and the lowland is almost free of them (fig. 79).

The climate is rainy and warm in summer and snowy and cold in winter. The precipitation, 600 to 800 millimeters per year, includes 130 millimeters of snow because of the high mountains. The temperature in the summer averages 30 C, while the low temperature in winter is about −8 degrees. The average annual temperature is 14 C. The region is extensively cultivated because the hard (Q1) loess soil lends itself well to agriculture.

Building materials are similar to those used in other regions: loess soil, sun-baked or oven-burned bricks, and sometimes stone. The area surrounding the cave-dwelling village is loess soil, terraced and leveled at approximately 100-meter intervals to overcome topographical variations. Under such environmental conditions many cliff cave dwellings, facing either south or east, have been dug into the terraces. The large courtyards are on the same level as their corresponding room units, and loess walls, three to four meters high, surround them on three sides (fig. 80).

Lu Chi is a typical village containing a large number of cliff cave dwellings (fig. 81). A dwelling complex for a family of six in Lu Chi consists of a set of rooms surrounding a courtyard or patio which is itself level with the road. There are two large rooms dug into each of three sides of the cliff. The center units facing east or southeast receive abundant sunlight, as do the units facing south. The third side, facing north, is shaded; and those rooms are used only for storage. The unit facing east is used as a kitchen, having an oven and storage space. All other rooms are used for living and contain some storage space as well. Each room has a door, a window, and a small opening at the top of the facade. The rooms are about 3.5 meters wide by 5 to 6 meters long by 3 meters high. A cistern stands in the middle of the patio to collect water. A low wall is placed about 3 meters inside the main entrance with the object of shielding the activities of the patio from the view of outsiders (fig. 82). (Sometimes such a wall is built 3 meters outside the entrance). This design for privacy appears repeatedly in the cave dwellings of the village.

The seat of local government, Qian Xian, is located in a mountainous area near the Qian Ling Tomb, a historic site dating to the Tang dynasty (A.D. 618–907). One typical cave dwelling in Qian Xian is that of the Zhong Houn Gi family, located about half a kilometer from the avenue of sculptures (life-size replicas of animals and human beings) leading to the tomb. The Zhong dwelling has two patios and is the first of several dwellings all de-

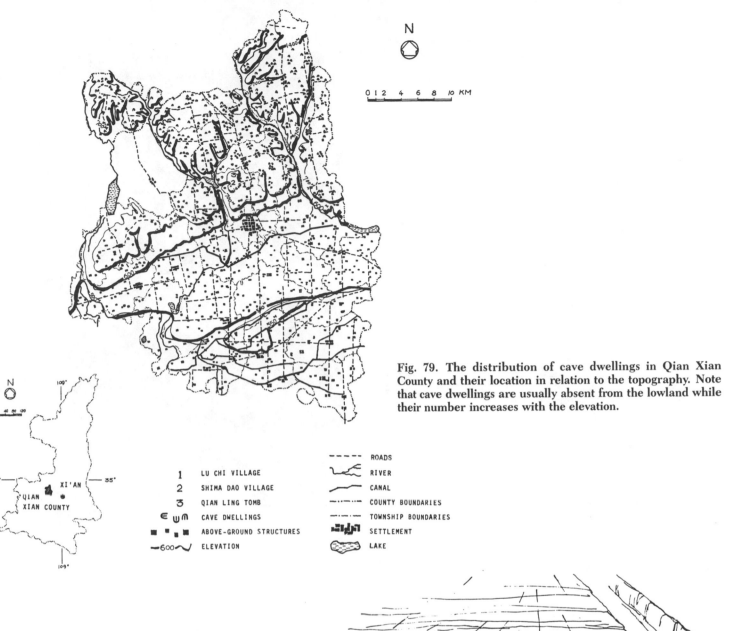

Fig. 79. The distribution of cave dwellings in Qian Xian County and their location in relation to the topography. Note that cave dwellings are usually absent from the lowland while their number increases with the elevation.

1   LU CHI VILLAGE
2   SHIMA DAO VILLAGE
3   QIAN LING TOMB
∈ ∪ ⋒   CAVE DWELLINGS
■ ▪ · ▪   ABOVE-GROUND STRUCTURES
∽600∿   ELEVATION

-----   ROADS
～   RIVER
—   CANAL
-·-·-   COUNTY BOUNDARIES
-··-··-   TOWNSHIP BOUNDARIES
⊡   SETTLEMENT
⬭   LAKE

Fig. 80. One type of cave dwelling in Lu Chi Village, Qian Xian County, Shaanxi province viewed from the east.

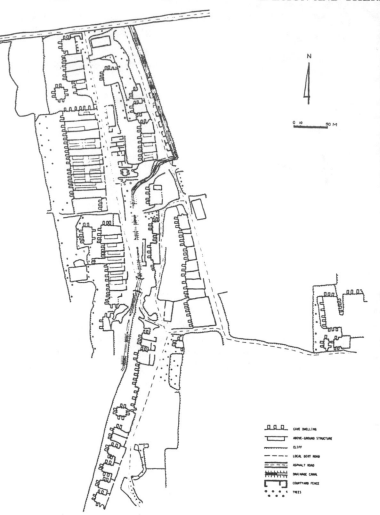

Fig. 81. Plan of Lu Chi Village cave dwellings. (See Fig. 79 for location.)

Fig. 82. Entrance to cliff cave dwelling in Lu Chi Village. The entrance road is level with the courtyard and the wall blocking a view of the interior.

signed in the same way. It is built on the same level as the dirt road, and room units are dug into a terraced cliff that forms three sides of a square (fig. 83). The fourth side, facing eastward toward the lowland, is enclosed by a wall 2 meters high. The house is well lighted and apparently well ventilated. In one room located on the north side, the ceiling was lowered by placing a beautiful bamboo grid covered with mate. The householder said that its purpose was not only to beautify, but also to prevent dust from falling from the ceiling into the room.

At a second cave dwelling nearby, the main entrance is blocked by an exterior wall. The entranceway itself is a below-ground corridor on the same level as the road and the courtyard. The courtyard is 12 meters square and contains many fruit trees. Below-ground dwelling units surround

Fig. 83. General view of cave dwelling community in Qian Xian County, Shaanxi province. Note the terraces and the construction of cave dwellings along the cliffs with patios level with the roads and the entrances.

this patio, two rooms on each side, with the entrance on the east. The cistern is located in a niche in the west wall between two cave units. The rooms are of standard size and shape: 4 by 7 by 3 meters high, with curved ceilings, a door, a window (1 meter square) set into the wall a little higher than the top of the door, and a small opening of 0.60 meters at the top of the facade to allow light to penetrate the interior. A typical room contains a heated bed, sewing machine, desk and chair, oriental closet (with an upper side opening), and many pictures. Eight persons live in this cave dwelling, which was built in 1963. The inhabitants like the dwelling very much, finding it cool in summer and warm in winter. They said the main problems were humidity in rainy weather, low light penetration, and soil moisture, the result of the relatively thin layer overhead (only 1.8 to 2 meters). They recognize now that they should have dug the dwelling deeper.

The village of Han Jai Po is located close to the foot of the mountains. Here, cave dwelling design is different. The cave dwelling is approached via a sloping path leading to the main entrance, which in turn leads to a little used courtyard, 10 meters square. In the middle of this courtyard are a cistern and a manual water pump and beyond a dividing wall there is an abandoned, collapsed cave dwelling. To the left, facing south, there is an entrance tunnel, 12 meters long and 3 meters wide, to the second courtyard. The inner courtyard is 8 by 12 meters with a tree in the middle and is surrounded by cave room units. The soil above the units is 2 to 3 meters thick and the rooms are of standard size. The unit facing west is used as a kitchen, while the unit with southern exposure is used as a living room and also for storage of wood and grain. The unit facing east, although it has a bed, does not seem to be in use, possibly because the ceiling is very damp. The room facing north is also standard size and not much used. The dwelling and courtyard are well lighted and proved an excellent environment for the four persons living there, the parents and two small children.

The following information was obtained from the mayor and applies to 1983.

Total population of Qian Xian County, 430,000
Land area, 181,170 acres
Number of settlements:
    Communes, 22
    Townships, 3
    Cities, 3
    Large teams, 317

Small teams, 1,920
Size of the largest team, 4,000–6,000
Size of the smallest team, 700–800
Qian Xian, the largest city, 30,000

The mayor also provided the following data about local cave dwellings:

Number of communes having cave dwellings, 9
Number of persons living in cave dwellings (i.e. more than one-third of the population), 150,000
Number of families denied cave-dwelling building permission in 1983, 300
Number of families given permission to build cave dwellings in 1983, 1,000
Number of persons who died as a result of cave-dwelling collapse (nearly unprecedented rainfall in 1983 caused the collapses), 90

BAI LESHENG DWELLING

The Bai Lesheng family pit cave dwelling is in Shima Dao Village in the northern part of the county. The environs are green and fully cultivated. The village is set at a high elevation and the cave dwellings are constructed in flat areas at the edge of a plateau. The road along the top of the ridge is the avenue of sculptures leading to the Qian Ling Tomb and overlooks the cave dwelling units under investigation. The view over the terraced valley to the west is very impressive. Nearby are many other pit cave dwellings on flat sites having their patios open to the sky. All the courtyards are level with the main entrances and with the dirt access road (fig. 84).

In principle, the overall design of the Bai Lesheng dwelling is very similar to the prototype Beijing vernacular above-ground house. It features two enclosed courtyards, one transitional to the inner, more private one (fig. 85). The main courtyard is 9 by 9 meters and is surrounded by cave units (fig. 86). The ceilings of the rooms slant downward at the rear and the thickness of the soil above the rooms is 4 to 5 meters (fig. 87). The small transitional entrance patio contains agricultural equipment. A below-ground passage, 3 meters long and 2 meters wide, connects the two courtyards, which are on the same level as the roadway to the west.

The inner courtyard has the entrance passageway and one dwelling unit on the westside. Two cave units are dug into each of the other three sides (fig.

Fig. 84. General view of Bai Lesheng family dwelling (No. 7), Shima Dao Village, Qian Xian County, Shaanxi province. The home (center) appears with its environs, from the west. Although the cave dwelling units here are the pit type, they are located on the terrace overlooking the lowland. At the top is the avenue of sculptures leading to Qian Ling Tomb.

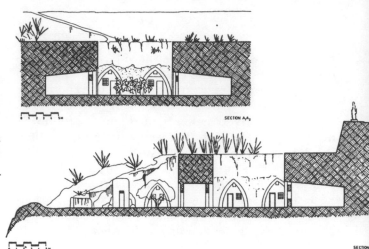

Fig. 87. North-south (A1, A2) and east-west (B1, B2) cross sections of Bai Lesheng family cave dwelling (No. 7).

Fig. 85. Perspective of Bai Lesheng family dwelling (No. 7).

Fig. 88. View inside the patio of Bai Lesheng family dwelling (No. 7).

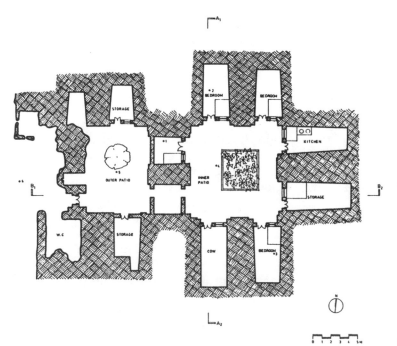

Fig. 86. Plan of Bai Lesheng family dwelling (No. 7).

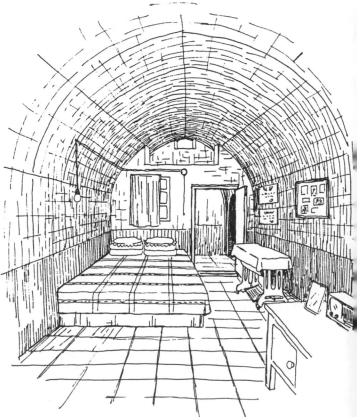

Fig. 89. Inside one of the cave dwelling rooms of Bai Lesheng family dwelling (No. 7).

88). The courtyard is planted with sunflowers for both utility and decoration. Most of the rooms are simply furnished and contain the typical heated bed (fig. 89). Three generations (eleven people including five children) share the dwelling. The northern room, used by the grandparents, is 6 by 3 by 3 meters high. One of the eastern room units is used for daily living and contains food storage jars. The other eastern room is without a door and is used simply for storage. The south has two rooms, one of which is lived in, while the other houses a cow and some straw. The owner of the house is a farmer who cultivates nearby land. The dwelling complex was built about twenty years ago as a group effort by the family.

THERMAL PERFORMANCE

At the Bai Lesheng family dwelling, six indoor and outdoor sites were measured; however, only four are discussed here (fig. 90). Sites 1 and 5 are represented by Sites 2 and 3 respectively. Summer dry- and wet-bulb temperatures of the bedroom (Site 2), which faces south, are different from those of the bedroom facing north (Site 3). In general, Site 2 is warmer than Site 3 over the twenty-four-

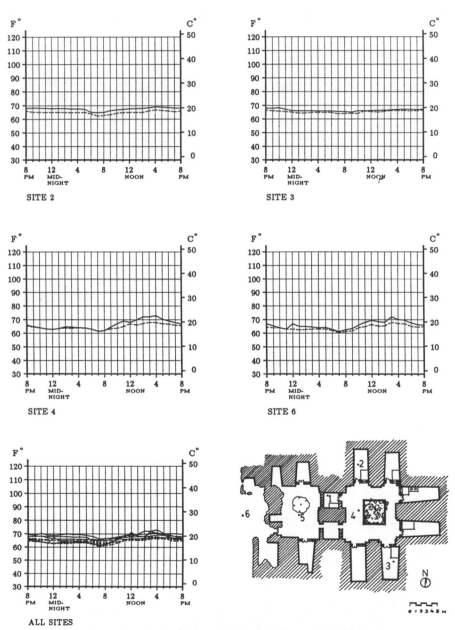

Fig. 90. Summer dry-bulb (solid line) and wet-bulb (broken line) diurnal temperatures, 27–28 August 1984, of Bai Lesheng family dwelling (No. 7).

hour period because of its orientation. The difference between daytime dry- and wet-bulb temperatures is greater in Site 2 than in Site 3. Site 2 temperatures are more stable during the night. During the day, temperatures at Site 2 fluctuate especially in the afternoon when they are warmed by the sun and reach a peak at 4:00 P.M. At Site 3, temperatures are stable and reasonably cool throughout the day and do not reach a peak point in the afternoon. Sites 2 and 3 show great contrast compared to the patio (Site 4) and with the outdoors (Site 6). The patio fluctuates during the daytime and

both temperatures reach a peak between 2:00 and 4:00 P.M. with the dry-bulb temperature registering 22 C at 4:00 P.M. At Site 4, nighttime dry- and wet-bulb temperatures are almost identical to one another. The outdoor Site 6 shows more fluctuation than does Site 4, especially in the afternoon hours. The dry-bulb temperature reaches its peak point of 22 C at 3:00 P.M. Its wet-bulb temperature parallels the dry-bulb temperature throughout the day but remains cooler. Temperatures of all sites indicate that the differentiation does not exceed 5 degrees, with the maximum at 22 C (dry-bulb) and the minimum at 17 C.

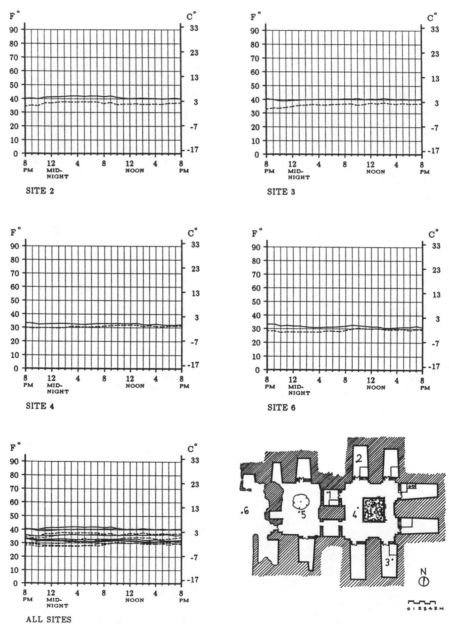

Fig. 91. Winter dry-bulb (solid line) and wet-bulb (broken line) diurnal temperatures, 7–8 December 1984, of Bai Lesheng family dwelling (No. 7).

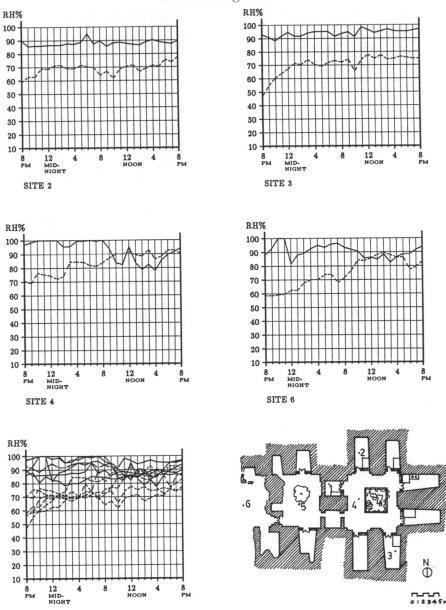

Fig. 92. Summer (solid line) and winter (broken line) relative humidity of four sites within Bai Lesheng family dwelling (No. 7).

Winter temperatures show much more diurnal stability at the four sites than do the summer ones (fig. 91). Maximum dry-bulb temperatures do not exceed 5 C (Site 2, bedroom) and the wet-bulb temperature does not go below the level of −3 C (Site 6, outdoors). Site 2 is warmer than Site 3, yet they both require some heating during the twenty-four-hour period. Obviously, the temperatures of the courtyard (Site 4) and of the outdoors (Site 6) are lower than those of Sites 2 and 3, yet they are both stable diurnally and close to one another. Site

4 is somewhat cooler during the twenty-four hours because of the lack of circulation within the patio itself, which retains stagnated air. Temperatures of all sites indicate more diurnal stability than those of the summer.

The relative humidity of Dwelling No. 7 indicates much differentiation between summer and winter (fig. 92). Because of the rain, summer humidity is very high and frequently reaches 90 percent or more at the four sites. The pattern of the two rooms (Sites 2 and 3) is similar but much different from the pattern of the patio and the outdoors (Sites 4 and 6). The afternoon heat of summer reduced the relative humidity and it became closer to, or even lower than, that of the winter because of the high elevation of the site.

Dwelling No. 8: Xing Xigeng Family Home, Gansu Province

ENVIRONS

The region of Qing Yang in the extreme eastern part of Gansu province, encompasses 27,000 square kilometers in seven counties. Most of the area is a plateau of medium-weight loess soil. It is green and well cultivated. However, erosion of the loess is intense because of the amount of rain. An old Chinese saying has it that the courtyard and the walls of a home keep in the wealth. However, the people of this region are poor and very isolated. Economic problems include the difficulty of marketing agricultural products (the nearest railway station is 400 kilometers away) and the lack of an advanced agrarian technology.

Below-ground dwellings constitute 83.4 percent of the total housing in Qing Yang, the highest percentage in Gansu province, and represent one of the densest concentrations of cave dwellings in China. The entire region exhibits various unusual designs of both the pit and the cliff types of subsurface dwellings (fig. 93). The town of Xifengzhen, where Dwelling No. 8 was researched, is a major center in the Qing Yang region (fig. 94). The cave dwellings of Xifengzhen are situated along one of the two main streets of the town.

This plateau region is the most humid in Gansu province since the elevation is high—between 1,200 and 2,000 meters above sea level—and annual precipitation averages 400 to 600 millimeters per year. The rains come at the end of July and extend into September. The annual snow accumulation is 20 centimeters, with the heaviest snows falling in January and February. The maximum temperature recorded in 1984 in Xifengzhen was 35.7 C, in July; the minimum was −22.6 C, registered in January.

XING XIGENG DWELLING

The Xing Xigeng family dwelling is a 150-year-old pit cave dwelling with a square patio, surrounded by many other pit cave dwellings and above-ground houses (fig. 95). A typical large family

Fig. 93. Courtyard of a pit cave dwelling, Xifengzhen, Qing Yang region, Gansu province.

cave complex of the Qing Yang region (fig. 96), entrance is gained by way of a descending graded ramp. Another saying heard in China is: "In the winter live in a cave dwelling and in the summer live in an above-ground house." During interviews with the residents, a young person in the family stated that he preferred the above-ground houses, while an elderly woman favored the cave dwelling if the summer humidity problem could be solved. The high summer humidity damages furniture and weakens the structure of the cave dwelling itself. Other complaints were that the residents must *descend* to enter their underground living quarters, causing unpleasant psychological associations; that the cave dwellings receive too little natural light; and that during rainstorms, they feared collapse of the dwellings. The high interior humidity is the result of a lack of ventilation common to all Chinese cave dwellings. Most residents have planted thickly branched fruit trees in their patios with consequent intense evapotranspiration, adding to the humidity problem. Accumulated rainwater at the center of the courtyard adds even more dampness to the atmosphere. The two families occupying Dwelling No. 8 have lived there for more than twenty years.

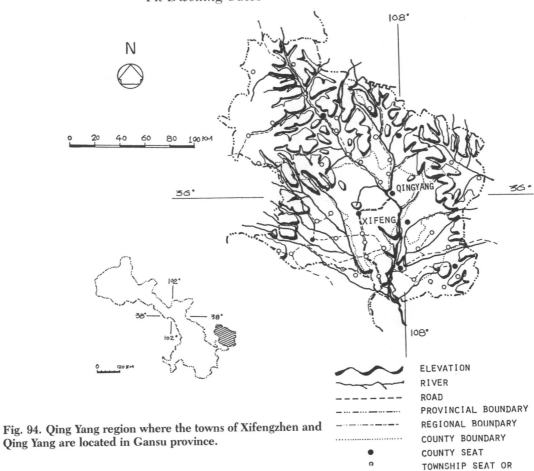

Fig. 94. Qing Yang region where the towns of Xifengzhen and Qing Yang are located in Gansu province.

Fig. 95. Xing Xigeng family pit cave dwelling (No. 8), Xifengzhen, Qing Yang region, Gansu province. The dwelling (on the right) is surrounded by many similar cave dwellings as well as above-ground houses.

Fig. 96. Perspective view of Xing Xigeng family dwelling (No. 8).

The larger Xing Xigeng family occupies the north, west, and south sides of the complex (fig. 97) and the other family occupies the east side.

Each dwelling unit has a mortared adobe facade with a brick foundation, 60–80 centimeters thick. The top of the door is level with the top of a small window (60 by 80 centimeters). Above this is a window (40 by 40 centimeters) for additional ventilation and light. In some cases there is still an-

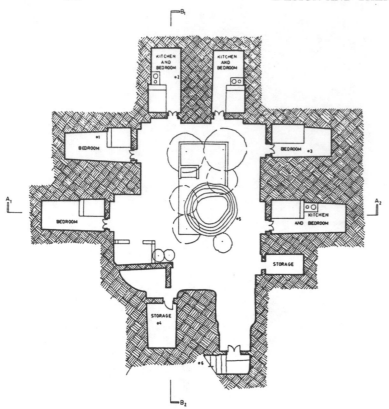

Fig. 97. Plan of Xing Xigeng family dwelling (No. 8).

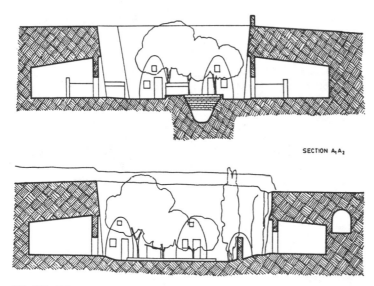

SECTION A₁A₂

SECTION B₁B₂

Fig. 98. North-south cross section of Xing Xigeng family dwelling (No. 8).

other opening (25 centimeters radius) at the very top. All the doors and window frames are painted black. The rooms are mortared and whitewashed

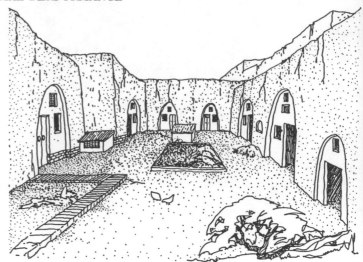

Fig. 99. The courtyard of the Xing Xigeng family dwelling (No. 8).

inside, the ceilings vaulted. The entrance to the complex is prudently placed in the south, which is the least desirable side to live on since it faces north and gets no sun. However, the summer prevailing winds come from the south and the entrance opening thus helps to ventilate the courtyard, which is 8 meters deep and shaded by the thick branches of an apple and a pear tree (fig. 98). Although providing shade, the fruit trees also increase humidity and minimize ventilation—two conditions not at all desirable in the summer season. In the middle of the patio there is a depression to collect rainwater (fig. 99).

THERMAL PERFORMANCE

The Xing Xigeng family dwelling is a pit cave dwelling typical of the Qing Yang region. Six sites were selected for temperature measurements. However, analyses of only four are included in this chapter.

Summer temperatures in Sites 1, 2, and 3 show very little diurnal fluctuation (fig. 100). The wet-bulb temperature, usually lower than the dry-bulb, is identical in pattern and parallel to it with little differentiation. The bedroom facing east (Site 1) has a stable temperature except at 11:00 A.M., when it reaches its peak two degrees above the average of 21 C. The bedroom/kitchen (Site 2) faces south. Both its diurnal dry- and wet-bulb temperatures are steady but higher than those of Site 1 because of the orientation and also because the residents are usually cooking there during the afternoon and eve-

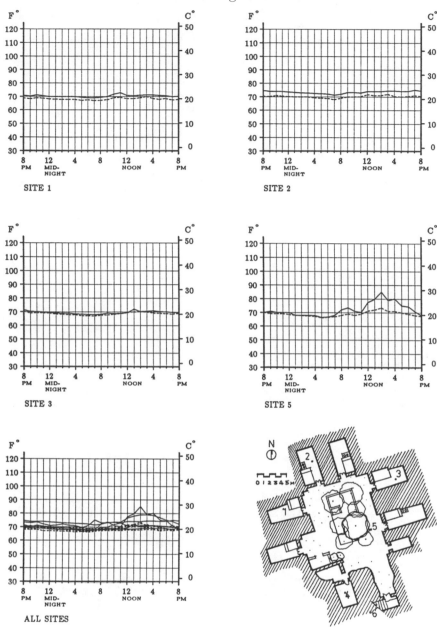

Fig. 100. Summer dry-bulb (solid line) and wet-bulb (broken line) diurnal temperatures, 31 July–1 August 1984, Xing Xigeng family dwelling (No. 8).

ning. In general, the temperature increase in Site 2 begins early in the morning. However, the highest point, 24 C, is reached at 7:00 P.M. At night, the dry-bulb temperature slowly and steadily drops to around 22 C. The room facing west (Site 3) follows a pattern similar to that of Site 2, yet it is used only as a bedroom. Temperatures decrease throughout the night to the lowest point of 18 C at 7:00 A.M. and begin to rise steadily to reach the highest point of 22 C at 2:00 P.M.

As is to be expected, the patio (Site 5) introduces different temperature patterns from those of the room units because of the cold, stagnated air. Here, too, the night temperature drops to its lowest point, 18 C, at 6:00 A.M. It rises sharply, fluctuating as it does, reaching a high point of 29 C (dry-bulb temperature) at 2:00 P.M. The wet-bulb temperature pattern is similar but not identical. Temperature graphs for Site 6, located above ground in the open space outside the complex, are not shown here because the pattern is very similar to that of the courtyard (Site 5). Except for the afternoon tem-

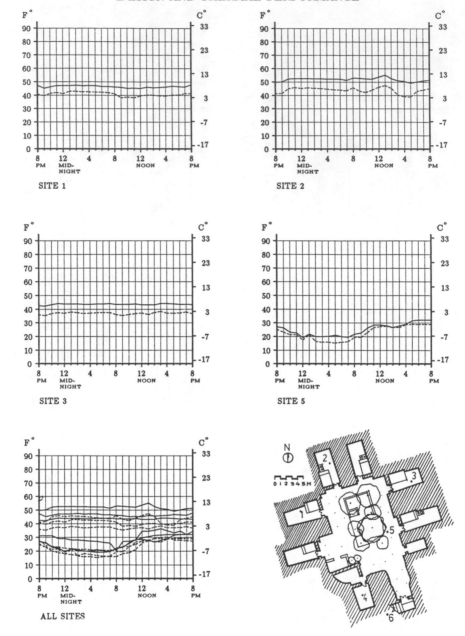

Fig. 101. Winter dry-bulb (solid line) and wet-bulb (broken line) diurnal temperatures, 5–6 December 1984, of the Xing Xigeng family dwelling (No. 8).

perature of the patio, the temperatures of all sites are grouped closely, between a maximum of 23 C and a minimum of 18 C, falling within the comfort range.

Winter temperatures in Dwelling No. 8 present quite a different picture from those of the summer (fig. 101). In general, temperatures of the three rooms do not go below zero and they reach a maximum of 12 C (dry-bulb) in the afternoon at Site 2 and at least 5 C (dry-bulb) at Site 3. In the three

rooms, both the dry- and wet-bulb temperatures run parallel to one another, with a differentiation of 4 or 5 degrees. The bedroom facing east (Site 1) has a steady dry-bulb temperature diurnally, with a minimum of 7 C in the early morning and at noon, and a maximum of 8 C throughout the rest of the day and night. The temperature fluctuates only slightly throughout the day and stays stable during the night. In any case, the room would require a slight amount of heating to bring it to the comfort

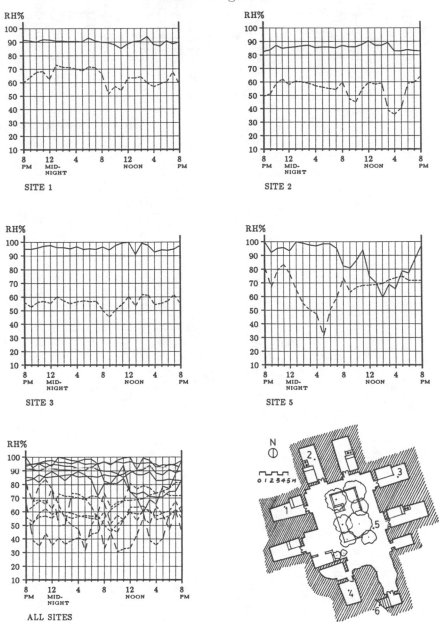

Fig. 102. Summer (solid line) and winter (broken line) relative humidity of Xing Xigeng family dwelling (No. 8).

zone. The bedroom/kitchen (Site 2) faces south and receives more natural heat than the other rooms. Consequently its afternoon temperature is higher: 12 C (dry-bulb) at 1:00 P.M. The lowest temperature is in the evening, 9 C (dry-bulb). The bedroom facing west (Site 3) has a stable diurnal temperature with little change in the evening. This site has the lowest average temperature among the three while Site 2 has the highest temperature. The courtyard (Site 5) has the lowest temperature among all the major sites, especially when it reaches −6 C at 7:00 A.M. Cold air stagnation in

the patio supports this development. This dry-bulb temperature is similar to that of Site 6 (outdoors in the open), which also reaches its lowest point of −6 C at 7:00 A.M. Thus, the pit cave dwelling patio has disadvantages in both summer and winter. The temperatures of all sites are grouped between the maximum of 12 C (dry-bulb) in the afternoon and −7 degrees C (dry-bulb) in the early morning, with a differentiation of 19 degrees.

The relative humidity of the summer (80 to 90 percent) is much higher than that of the winter (50 to 70 percent), as shown in fig. 102. Also, the summer measurements are very stable except in the courtyard in the afternoon, while the winter readings show a wide range of fluctuation. The most winter fluctuation occurs in the courtyard (between a low of 30 percent and a high 85 percent) because of the diurnal winter wind in this area. The drop in relative humidity during the summer afternoon, to below that of the winter, is the only such case occurring during the two seasons monitored, and it was due to the increase in the afternoon heat (dry-bulb). The entranceway on the south side of the courtyard supported some ventilation there.

Dwelling No. 9: Tian Lu Family Home, Henan Province

REGIONAL CHARACTERISTICS

Henan is the most densely populated of the provinces researched. The majority of the cave dwellings are concentrated in the rolling hills and mountains of the western part where the topography supports the construction of both the cliff and pit types. Most of the loess soil in the region is soft Ma Lan type (Q3) that requires special considerations in cave dwelling design. Bricks are the primary building material used in the area.

The average annual temperature of the region falls between 14 and 28 C, with the average in January being around 0 C and in July around 28 C. The minimum in January is −8 C and the maximum around 8 C. The minimum in July is around

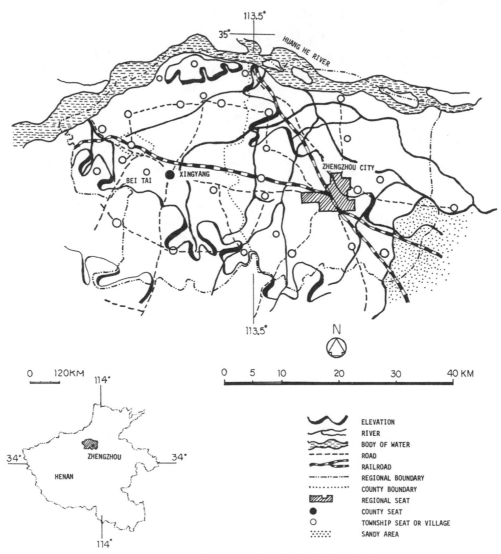

Fig. 103. Xing Yang County and Bei Tai Village in Henan province where the Tian Lu family dwelling (No. 9) is located.

18 C and the maximum, 35 C.

The annual precipitation is 400 to 600 millimeters, most of it during the period July through September, with occasional snowfalls in January and February. The relatively high amount of precipitation increases the moisture within the dwellings and introduces the threat of collapse, especially when the rain is heavy.

TIAN LU DWELLING

The village of Bei Tai is located in Xing Yang County, some 30 kilometers west of Zhengzhou, capital of Henan (fig. 103). The village is near enough to the city to feel its influence. The Tian Lu family dwelling is near a lake in a beautiful rolling environment (fig. 104), surrounded by other cave dwellings and above-ground houses. The Architectural Academy of Zhengzhou City has done some preliminary research and mapping of this dwelling. The interesting overall design of the complex was made by Tian Lu himself, who put much time and thought into it (fig. 105). In building his dwelling, Tian Lu expanded upon the Chinese tradition by providing two additional courtyards—the first, a large entrance area that is transitional to the main courtyard, and the other, a yard to house his livestock. As a result, he was able to have rear windows in two of the cave room units, thus providing much improved lighting and ventilation for the rooms as well as some air circulation in the main courtyard. The complex is occupied by a total of eleven per-

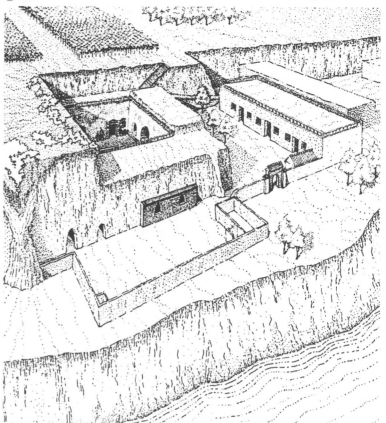

Fig. 105. Perspective view from southwest of Tian Lu family dwelling (No. 9). Note the three patios on the same level, the great depth of the central patio, and the agricultural crops on the surface. The entrance to the main patio (center) is from a transitional one on the right. The patio in front is for livestock.

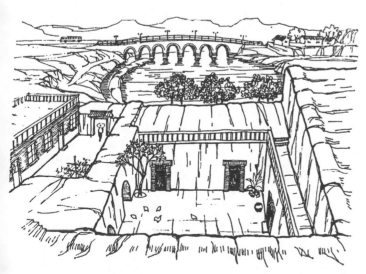

Fig. 104. View from the north: Tian Lu family dwelling (No. 9) and environs, Bei Tai Village, Henan province. Note the nearby lake and the above-ground house.

sons, all members of the Tian Lu extended family.

The three patios, the cave dwelling units, and the above-ground rooms are all on the same level. The entrance to the transitional patio is on the south side. Several above-ground bedrooms line and eastern wall of this patio. In the northwest corner is the below-ground entrance to the main central courtyard of the complex. The "barnyard" patio contains pens and caves for the livestock, a storage area, the privy, and an underground passageway northward to the kitchen.

The main feature of the cave dwelling complex is a large courtyard (9.5 by 7 meters) surrounded by room units that protect it from wind and snow (fig. 106). On the north side are two rooms facing south, one of which is used as a bedroom. On the west, one room is used as a kitchen. From it a narrow, curving tunnel leads to the animal quarters (figs. 107 and 108). On the south side there are two rooms, each with a window at the back overlooking

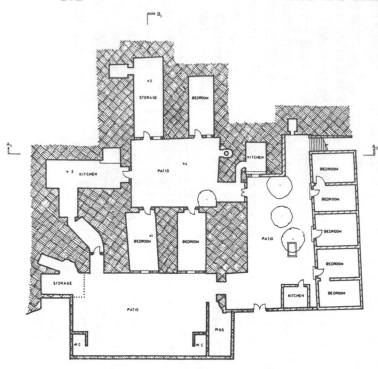

Fig. 106. Plan of Tian Lu family dwelling (No. 9).

Fig. 108. View of the kitchen (Site 2) of Tian Lu family dwelling (No. 9). Note the passage on the right to a storage area and the animal quarters. There is also a storage niche at the left side near the entrance.

Fig. 109. View of Tian Lu family dwelling (No. 9). Note the cliff at the rear and sides.

Fig. 107. View from the east to the main central patio of Tian Lu family dwelling (No. 9).

the livestock patio. These two rooms, with doors on one side and windows on the other, permit ventilation and air circulation between the two patios. Just inside the entrance to the main patio is a small corridor giving access to a second below-ground kitchen. At the northeastern corner of the central patio is a niche, elevated above the patio floor to avert rainwater, for storage and preservation of fruit and vegetables in relative coolness. The complex is quite unusual, but the great depth of the patio creates a serious problem with thermal performance by diminishing ventilation (fig. 109). Having the eastern side of the patio cliff lowered to 8 meters increases light and sunshine penetration (fig. 110).

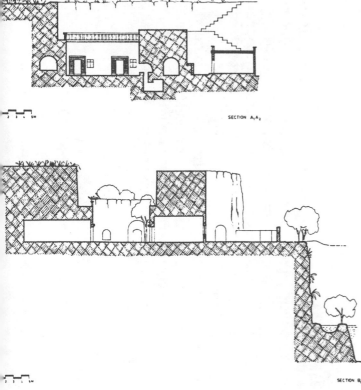

SECTION A,A,

SECTION B,B

Fig. 110. Cross sections of Tian Lu family dwelling (No. 9).

The heat generated at noon within the main courtyard resulted in some of the highest temperature readings we encountered in the twelve cave dwellings analyzed, reaching 43 C by 11:00 A.M. Like the other pit cave dwellings in China, this complex is deficient in sunshine, light, and ventilation. In addition, like the other pit cave dwellings in Henan province, it suffers from high indoor moisture and humidity.

The cave dwelling units have 10 meters of terraced cliff above the north side, with a low brick wall around the perimeter of the patio opening. The soil above the dwellings is cultivated. The Tian Lu family, in contrast to most Chinese, separates the room functions within the cave dwelling and also separates the animals from the main household courtyard.

THERMAL PERFORMANCE

From the thermal point of view, the uniqueness of this pit cave dwelling complex is that the depth of the main patio exceeds 10 meters on two sides; it has cross ventilation in three rooms, enabling air to circulate, and it is located near a body of water. All

of these make this dwelling different from the other pit cave dwellings researched.

Summer temperatures were measured in five indoor and outdoor locations. However, only four are shown on the graphs (fig. 111). The kitchen (Site 2) has a stove that was used during some parts of the day but not at night. Dry-bulb and wet-bulb temperatures are almost identical diurnally with a differentiation ranging between 1 and 3.5 C. Night dry-bulb temperature is almost stable and decreases by 1 degree toward the early morning hours and rises during the day to a peak of 29 C at 1:00 P.M. The difference between the lowest dry-bulb temperature at night and the highest dry-bulb temperature in the afternoon is 4 degrees C; both temperatures stay within the range of relative comfort. The storage room (Site 3), located on the northeast side of the main patio, did not have ventilation, yet it was cooler than the kitchen (Site 2), doubtless because of the stove operating there. Its temperature was within the range of comfort: a maximum of 24 C (afternoon) and a minimum of 22 C (night). The wet- and dry-bulb temperatures are almost identical diurnally with only an occasional small gap. It would seem that the dampness in Site 3 has affected the temperature causing it to be different from that of Site 2.

The central patio (Site 4) is basically similar to other pit house patios; however, it is different in the early temperature peak, 43 C at 11:00 A.M., while the wet-bulb temperature at 11:00 A.M. was only 26 C. This differentiation and early peak is due to the great depth of the central patio in relation to its width and length and to the lack of ventilation. We believe that the owner opened an exit from the kitchen (Site 2) to the south patio to increase the air circulation of the central patio (Site 4) when he realized how intense the heat was there at noon in summer. The outdoor temperature (Site 5) has three peaks during the day (at 11:00 A.M. and at 1:00 and 3:00 P.M.). Still, the temperature there is lower than that of the patio, and at 3:00 P.M. did not exceed 35 C. In general it was more comfortable to be outdoors than to be in the patio during the daytime. Temperatures of all sites show a range of 6 C between maximum and minimum, with the exception of the dry-bulb temperatures of the patio and of the outdoors from 11:00 A.M. to the evening hours.

Winter temperatures also show great differentiation between indoors and outdoors (fig. 112). The dry-bulb temperature measurements of Sites 2 and 3 never dropped to the freezing point; the wet-bulb temperature did so throughout the night at Site 3.

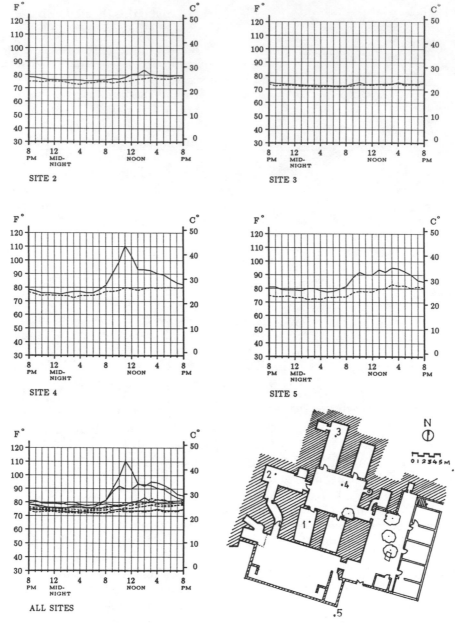

Fig. 111. Summer dry-bulb (solid line) and wet-bulb (broken line) diurnal temperatures, 4–5 August 1984, of the Tian Lu family dwelling (No. 9).

With regard to the patio (Site 4) and the outdoors (Site 5), both dry- and wet-bulb temperatures dropped below freezing during the entire night and much of the day, thus showing the advantages of the below-ground space. However, both Sites 2 and 3 necessitate some heating throughout the twenty-four-hour period to bring temperatures to a comfortable level. The temperatures in the kitchen (Site 2) are higher than those at Site 3 because of the use of the stove in the kitchen, yet the overall tem-

perature pattern of the two rooms is very similar. The outdoor temperatures (Site 5) are similar at night to those of the patio (Site 4), but they are higher in the afternoon hours. Thus, here too it is more comfortable to be out in the open than to be in the patio during the afternoon. Among all the sites measured in this complex, the worst one was the patio in the afternoons both in summer and in winter. The temperatures of all sites have a wide range with the maximum of 7 C and the minimum

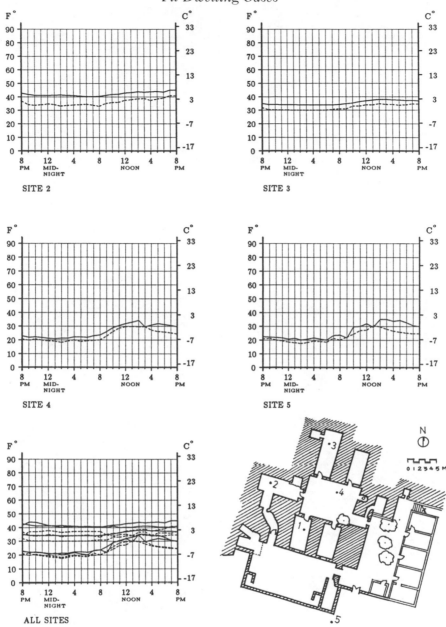

Fig. 112. Winter dry-bulb (solid line) and wet-bulb (broken line) diurnal temperatures, 24–25 December 1984, of the Tian Lu family dwelling (No. 9).

dropping to −9 C, wet-bulb temperature.

The relative humidity of the two rooms (Sites 2 and 3) are similar but not identical (fig 113). In the kitchen (Site 2) much diurnal fluctuation is shown in both seasons (due to the use of the stove and the ventilation), with a great gap between the maximum and the minimum humidity, especially at night: summer, 96 percent; and winter, 43 percent. At Site 3, the north room, both summer and winter humidity fluctuate less, and there is a smaller gap

between the maximum of summer (97 percent) and the minimum of winter (67 percent). Both the central patio (Site 4) and the outdoors (Site 6) show great fluctuation in the summer as well as in the winter. The summer afternoon relative humidity drops to 29 percent due to intense heat evaporation, while that of the winter afternoon increases to 83 percent. A similar interchange occurs outdoors (Site 5) diurnally.

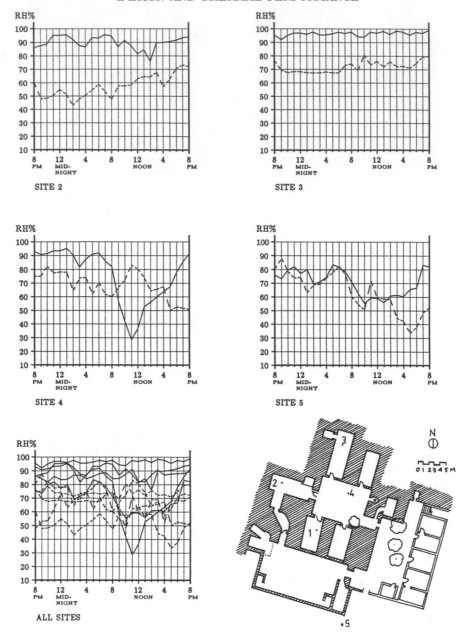

Fig. 113. Summer (solid line) and winter (broken line) relative humidity of Tian Lu family dwelling (No. 9).

Dwelling No. 10: Yin Xin Yin Family Home, Henan Province

ENVIRONS

Gong Xian County is located south of the Huang He River, 50 kilometers east of the city of Luoyang and 75 kilometers west of Zhengzhou, the capital of Henan province. The county is agricultural with a number of historical sites. The total population is 600,000 people, of which one-third live in cave villages and towns. Most of the dwellings are the pit cave type. Two cave dwellings were researched in this county, Dwelling No. 10, in Xi Cun Village, and No. 11 in Gong Xian Town (fig. 114).

The topography of the county is mostly rolling with elevations ranging from 200 to 2000 meters. The Yiluo He River winds through the region. In the south is Song Mountain, while in the west is the historic Heishiguan Pass, and to the east is the

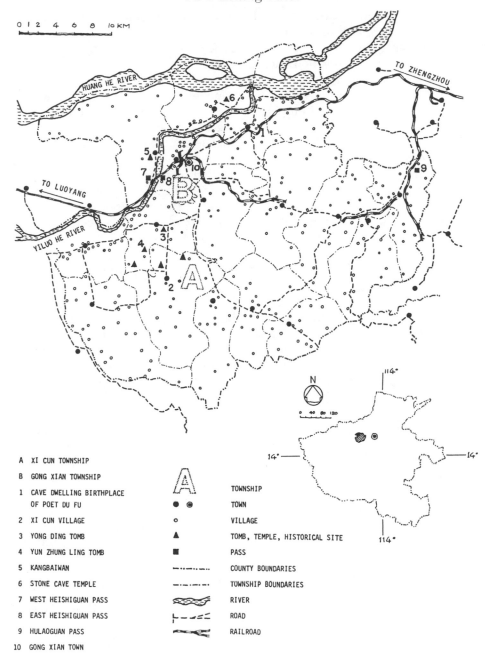

Fig. 114. Gong Xian County and the location of Xi Cun Village, Xi Cun Township (Dwelling No. 10), and Gong Xian Town (Dwelling No. 11), Henan province.

Hulaoguan Pass. The soil is Ma Lan type (Q3), which is soft loess.

The climate is moderate, that is, rainy and warm in the summer (primarily July and August) and snowy and cold in the winter (December through February). According to the local people, there is a heavy snow only every few years. The average annual precipitation is 600 millimeters, with this re-

gion receiving higher summer precipitation than any of the other four provinces researched. The combination of high summer heat and humidity causes serious problems for the cave dwellings in Henan. The average annual temperature range in this area is from 8 to 14 C. In five months, May through September, the average temperature exceeds 28 C. From November to February the temperature hovers around the freezing point. Optimal conditions for usage of the cave dwellings primarily occur during the winter season.

Gong Xian County has a strategic position and is regarded as the key to Luoyang City, an eastern capital of ancient China. The county's history can be traced back to 221 B.C. under the first emperor of the Qin dynasty. Nearby is Nanyaowan, the birthplace of the famous realist poet, Du Fu (A.D. 712–770), of the Tang dynasty. His poems had a strong influence on later generations, and he recorded the history of his era. The region has other important historical sites such as the Stone Cave Temple, the ancient kiln where Tang tri-colored glazed pottery was made, the Song Ling Tombs of the Northern Song Dynasty, and the architecturally interesting buildings of the Manor of Landlord Kang Baiwan.

The Stone Cave Temple consists of a series of Buddhist caves located 10 kilometers north of the town of Gong Xian at the foot of Dali Mountain overlooking the Yiluo He River valley. The temple was first built in A.D. 517 and later enlarged. The caves are carved out of limestone below the layer of loess soil. One cave contains almost one thousand tiny sculptures of Buddha.

The Yong Ding Tomb of the Song Emperor, like many other ancient Chinese tombs, is below-ground. An avenue, marked by two rows of larger-than-life sculptures of human beings, elephants, tigers, and lions leads to the hill containing the earth-covered tomb. Another tomb, the Yung Zhung Lin, has the same style of avenue of large sculptures leading to an artificial hill burial site.

The Landlord Kang Baiwan complex was built about 300 years ago with the cliff caves on the same level as the other above-ground structures. We estimated the cliff to be more than 20 meters high, there being more than 15 meters of soil above the cave units. Entrance is gained through large below-ground tunnels. Two steps lead up from the patio to each cave, which is two stories high with high ceilings as well. The cliff walls outside the caves are covered with brick to more than 3 meters above the ceiling level of the rooms.

The poet Du Fu was born in a cave at the foot of the Bijiashan Mountain, Nanyaowan. The dwelling is located approximately 15 kilometers northeast of the town of Gong Xian. In front of the cliff cave dwelling there is a large patio bounded by a building on the west, another building and a wall on the south, a wall on the north, with the cave itself on the east. The cave is 8 to 9 meters long, 3 to 3.5 meters wide and 3.5 meters high. The vault, the

walls and the floor are covered with old gray brick.

THE VILLAGE AND THE DWELLING

North of the Yiluo He River, there are large concentrations of cave dwellings including a few ancient ones. The village of Xi Cun, where Dwelling No. 10 is located, is in Xi Cun Township in the southern part of the county. Xi Cun is a rather large agricultural commune. Pit cave dwellings prevail as the land is almost flat (fig. 115). However, some new above-ground structures are scattered around the village. This flat area is intersected by a gully and some dwellings have been built in its sides. Xi Cun Village has a population of 6,000 people, and half live in caves. The entrance to the dwellings is often via a stairway paved with locally made bricks or stones from the nearby mountains. In some cases, the entrance walls are attractively lined with burned bricks. Every entrance has one or two ninety-degree turns before one enters the patio. In this way, there is no direct view inside, although the pit courtyard can easily be viewed from above by a stranger. Most patios are square (8 by 8 or 10 by 10 meters). However, some are rectangular with a length of 12 meters or more. Depth is usually 8 to 10 meters, leaving 3 to 5 meters of earth above each dwelling. The trees planted within the courtyards often occupy much of the pit space and tend to limit ventilation that is badly needed becuase of the high humidity in the rooms in the rainy season. Some dwellings have two courtyards, a small transitional one and the main central courtyard. In some cases, the passage from the first to the second is by way of a below-ground tunnel, 4 to 6 meters in length.

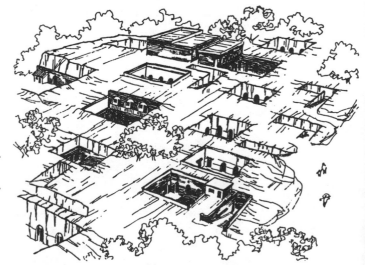

Fig. 115. Section of Xi Cun Village where Yin Xin Yin family cave dwelling (No. 10) is located (lower right).

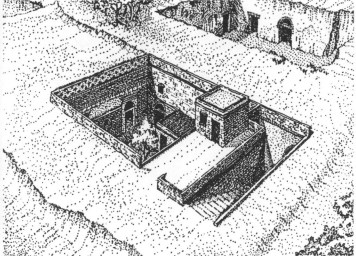

Fig. 116. Bird's-eye view of Yin Xin Yin family dwelling (No. 10), Xi Cun Township, Gong Xian County, Henan province.

SECTION A₁A₂

SECTION B₁B₂

0 1 2 3 4 5M

Fig. 118. Cross sections of Yin Xin Yin family dwelling (No. 10).

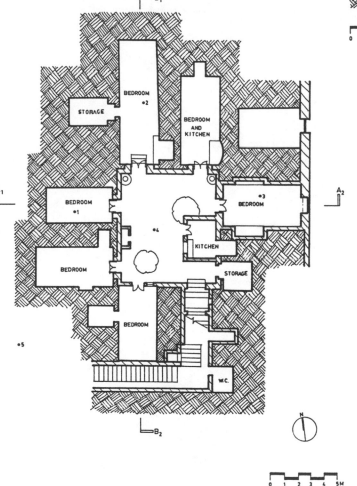

Fig. 117. Plan of Yin Xin Yin family dwelling (No. 10).

Fig. 119. The flight of stairs leading into the courtyard of Yin Xin Yin pit cave dwelling (No. 10). The privy is at the upper left.

Beside the entrance is often an open drainage ditch. Livestock is sometimes quartered in the outer patio.

Dwelling No. 10 is occupied by the Yin Xin Yin extended family, consisting of nine people. Like others in the area, this dwelling has an almost square pit patio (8 by 7 meters) surrounded by a brick wall on top (fig. 116). The design of the dwelling is typical of this village: a central courtyard surrounded by rooms on all four sides (fig. 117). The patio contains an above-ground kitchen (2 by 3 meters). The stairway and one bedroom are located on the south side. The north side contains two bedrooms (one of which is huge and has been designated Site 2) plus a storage room. One room on the east (Site 3) has extra flooring near the ceiling, creating a storage loft 2 meters high. All rooms are whitewashed to reflect the light. The patio walls are covered with brick for 6 meters (fig. 118). The entrance stairway is partly open to the sky and partly covered by earth (fig. 119). The present people are not the first owners, as the caves were constructed before 1940 and have been in continuous use ever since.

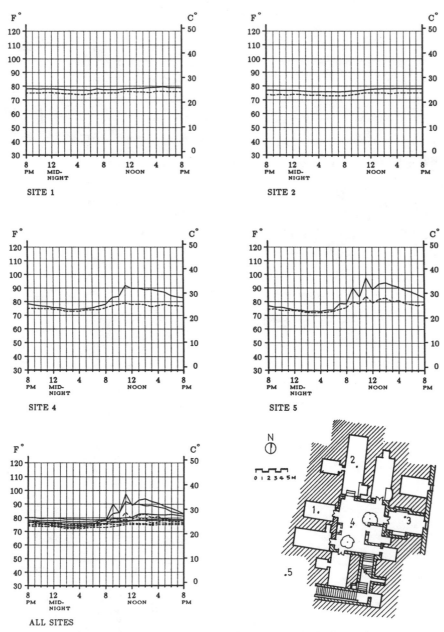

Fig. 120. Summer dry-bulb (solid line) and wet-bulb (broken line) diurnal temperatures, 7–8 August 1984, of the Yin Xin Yin family dwelling (No. 10).

THERMAL PERFORMANCE

Dwelling No. 10 is a pit cave dwelling that has rooms on all four sides of the courtyard. One room is longer than the length of the courtyard itself. Dry- and wet-bulb temperature measurements were made diurnally at five indoor and outdoor sites. (Only four are displayed in the graphs).

Summer temperatures of the two bedrooms (Sites 1 and 2) are almost identical to one another and are very different from those of Sites 4 and 5,

the courtyard and the outdoors (fig. 120). The bedroom (Site 1) faces east and therefore is well lighted and sunny in the morning. Both the dry- and wet-bulb temperatures are almost stable diurnally, with less than 2 degrees C between the maximum, reached at 4:00 or 5:00 P.M., and the minimum reached at 4:00 or 5:00 A.M. The long bedroom facing south (Site 2), although similar in temperature pattern to Site 1, is cooler both in summer and in winter. We credit this coolness to its size—more than 9 meters long—and believe that a

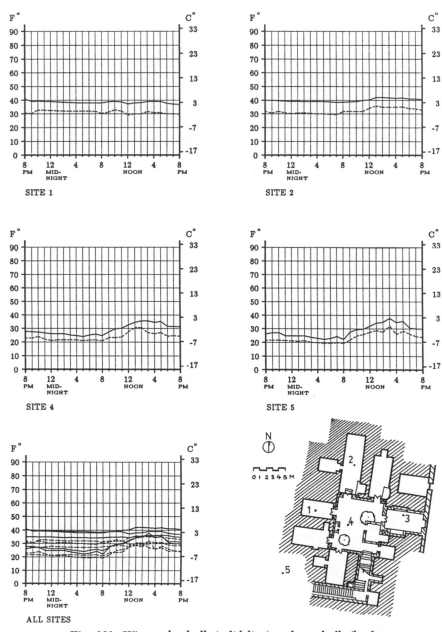

Fig. 121. Winter dry-bulb (solid line) and wet-bulb (broken line) diurnal temperatures, 28–29 December 1984, of the Yin Xin Yin family dwelling (No. 10).

lengthy room contributes to better thermal performance throughout the year because of the increased wall surface. In Site 2, the temperatures are almost parallel, with a difference of only 2 degrees between them. The afternoon temperature holds steady from 1:00 P.M. until the late evening. The dry-bulb temperature of the courtyard (Site 4) fluctuates and reaches a peak at 11:00 A.M. and then continues steadily downward until evening. There is quite a difference between the dry- and wet-bulb temperatures especially in the afternoon. However, the dimensions of the patio are small (7.3 meters wide by 8.3 meters long by 8 meters deep), which limits the movement of air and the possibilities for ventilation. Also, the patio is obstructed by an above-ground structure used as a kitchen. The drop in temperature in the patio late at night is sharper than that of the bedrooms because of the stagnated air there. Yet temperatures are higher than those of the outdoors (Site 5) at the same hour. The outdoor dry- and wet-bulb temperatures reach three peaks during the day: at 9:00 and 11:00 A.M. and at 2:00 P.M., and both drop steadily after 2:00 P.M. The dry-bulb temperatures of both the patio and the outdoors are high and uncomfortable in the afternoon and therefore they are quite different from those of the two bedrooms. The temperatures of all sites (with the exception of the very high afternoon temperatures of the patio and of the outdoors) are either within the comfort zone or just bordering it, with the highest dry-bulb being 26 C and the lowest wet-bulb being 22 C.

Winter temperatures introduce a different pattern (fig. 121). The two bedrooms (Sites 1 and 2) have generally stable temperatures with some minor fluctuations in the afternoon. At Site 1 the temperature difference between dry- and wet-bulb is 3 to 4 degrees and at Site 2 is 3 to 5 degrees. In Site 1, the maximum dry-bulb temperature is slightly higher than 3 C, while the wet-bulb temperature is slightly below freezing much of the time. A similar pattern appears in Site 2 where the maximum does not exceed 5 C (dry-bulb) in the afternoon and the wet-bulb temperature is slightly below freezing late at night. The courtyard (Site 4) is similar to the outdoors (Site 5). However, it is cooler in the afternoon with a maximum measurement of 2 C at 3:00 P.M. At the same hour the highest point of the outdoors (Site 5) dry-bulb temperature is reached, yet most of the time both sites are below the freezing point. Here too, it is more comfortable and relatively warmer outdoors than on the patio during the afternoon. As is the case with many other pit cave dwellings, the patio has serious

deficiencies at night as well as in the daytime. The combined temperatures of all sites are between 5 C (dry-bulb) and −7 C (wet-bulb). The two rooms (Sites 1 and 2), however, would require some heating throughout the twenty-four hours.

The relative humidity pattern is typical of this type of pit cave dwelling (fig. 122). There is a great gap between the high summer relative humidity (around 90 percent) and the winter relative humidity (45 to 55 percent) in Sites 1 and 2. The courtyard (Site 4) and the outdoors (Site 5) are similar to one another and quite different from the pattern of Sites 1 and 2. During the night and through most of the first half of the day, there is a large gap between the summer and the winter relative humidity at Sites 4 and 5. Proximity and interchange occur when the summer relative humidity drops from 90 percent to around 60 percent due to the intense heat in the afternoon.

Dwelling No. 11: Li Songbin Family Home, Henan Province

THE TOWN AND THE DWELLING

The town of Gong Xian, the seat of Gong Xian County, is located along a major road running between Zhengzhou, capital of Henan province, on the east and Luoyang, a major city to the west (fig. 114). A good-sized portion of the town is constructed below ground. The Li Songbin family cave dwelling is located within a below-ground neighborhood on the north side of the town (fig. 123). It was built in 1970 and is interesting because of the innovations made by the owner.

The soil atop the dwelling is on the same level as the street. When approaching from the south (the street side), the visitor must descend at least 11 meters to the below-ground alley and entrance (fig. 124). On the right side of the alley as one proceeds northward, there are entrances to many cave-dwelling complexes; on the left are the rear walls of other such dwellings. Li Songbin's is the first entrance along this alley. Above part of his southern cave units there is a carpenter shop whose machinery creates vibrations. Luckily, the dwelling does not have many living units on that side.

The dwelling has a huge rectangular brick-walled courtyard on the same level as the alley (fig. 125) with room units surrounding the patio on all sides (fig. 126). Close to the entrance to one room facing south (Site 2), Li Songbin dug a 5 meter shaft from the ceiling to the surface (fig. 127). He then added a

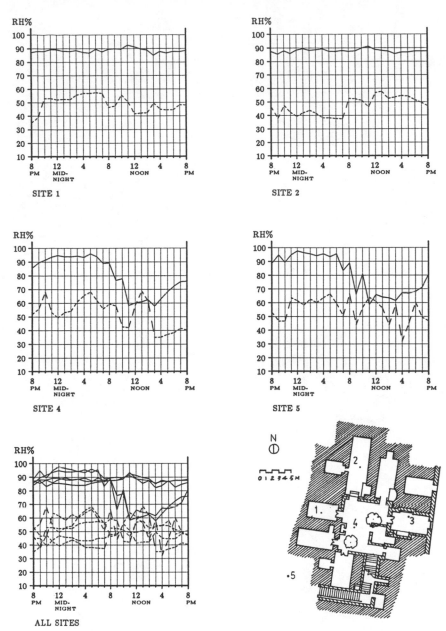

RH%
SITE 1

RH%
SITE 2

RH%
SITE 4

RH%
SITE 5

RH%
ALL SITES

N

0 1 2 3 4 5 M

Fig. 122. Summer (solid line) and winter (broken line) relative humidity in Yin Xin Yin family dwelling (No. 10).

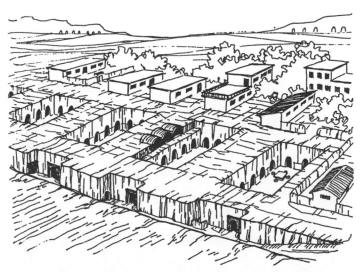

Fig. 123. The environs of Li Songbin family dwelling (No. 11; extreme right), Gong Xian Town, Henan province. Note the second floor addition of an earth-sheltered habitat (center) and overall vaulted design of the three rooms.

Fig. 124. Bird's-eye view of Li Songbin family dwelling (No. 11).

Fig. 125. Below-ground alley of the Li Songbin family dwelling (No. 11) and environs. The alley is level with the patio floor.

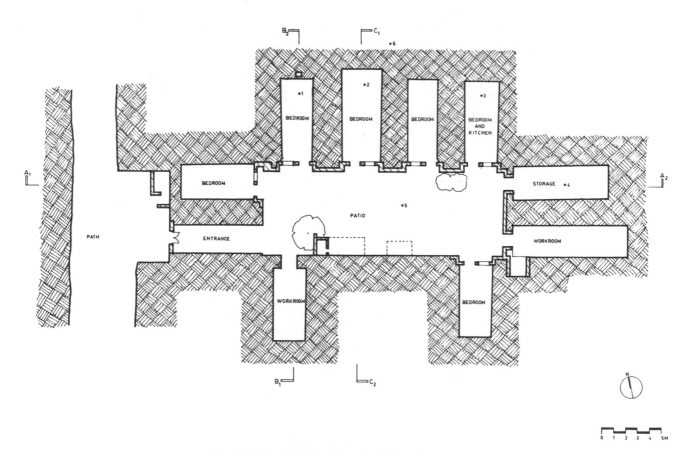

Fig. 126. Plan of Li Songbin family dwelling (No. 11).

plastic dropped ceiling half a meter lower than the actual ceiling, forming a tunnel open only at the end of the room—an attempt to create air circulation from the doorway to the end of the room and then back above the plastic all the way to the shaft. The room is about 7.5 meters long and 3 meters wide. In an adjacent room (Site 1), the owner reversed his design, starting at the rear with the shaft, and eliminated the need for the plastic layer. Neither of the shafts created air circulation, perhaps because of the small diameter of the air shafts. The owner-builder is participating with the cave-dwelling research group of the county, a part of the Henan province research group, in testing various projects.

The four main rooms are located on the north side (facing south) where Sites 1, 2, and 3 were selected for research. On the east side are two room units, one of them designated as Site 4. On the west there is one room unit along with the main entrance to the dwelling complex. On the south side are two units, one at each end. The entrances to Sites 1, 2, and 3 are screened to repulse mosquitoes. Sites 1 and 2 were selected for analysis because of their ventilation chimneys. Site 3, without a chimney, was chosen for comparison.

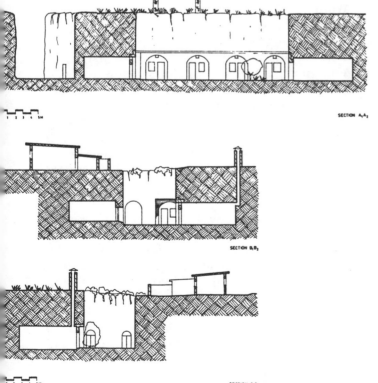

Fig. 127. Cross sections of Li Songbin family dwelling (No. 11).

Fig. 128. View of the patio of Li Songbin family dwelling (No. 11).

The bricks of the patio walls seem to be recently added, perhaps only a few years ago. Except for the south wall, the bricks cover to the top of the cave dwellings, a distance of 3 to 3.5 meters. Above this, there are approximately 5 meters of soil. On the north rim is an overhang to protect against rain (fig. 128). On the surface there are shrubs, and some of the soil above the dwelling is used for agriculture.

Because of the water tap in the patio, there is a drainage problem with water accumulation, mud, and mosquitoes. The typical rural privy is located outside near the dwelling entrance with an open sewer nearby.

Site 1 is approximately 6 by 3 by 3 meters high. The room is used as a living room/bedroom and as a storage area. The front part is shaded all afternoon. The floor is on the same level as the patio. Paper covers the walls to a height of 1.5 meters; above this level, the vault is nicely mortared and whitewashed. The room has a glass window that was closed throughout the time of the research. The upper part of the door also has a glass window that was also closed. At the top of the facade is a small opening 30 centimeters square. The outside brick vault design is a little larger than the actual inside vault. Site 1 is used as both a bedroom and a storage area. The experimental chimney begins at the top of the inside wall. The chimney opening is 37 by 25 centimeters, and the stack rises 2 meters above the soil surface. The chimney has a brick facing and a cover to deflect rain. The dwellers said that the chimney is not effective because it is too narrow. No movement of air could be felt in Sites 1 or 2.

Site 2 is used as a bedroom, and has the same

window and door design as Site 1. Here, also, the front wall is made of brick. This room, however, is wider and longer than Site 1: approximately 4 by 8 by 3.5 meters high. The chimney is at the front and seems to start at the ceiling, not from the wall. The tenant who designed the chimney seems to have revised his thinking after building the shaft and wanted air movement to go all the way along the room. To do so, he built a false ceiling of paper half a meter lower than the original ceiling, from the front of the room to one-half or three-quarters of a meter short of the back wall. According to Li Song-bin, the air should enter the room through the door or the window, move through the room, rise to the top of the vault at the rear, enter the passage created between the ceiling and the paper and then exit via the chimney. To prove his theory, he claimed that when smoke entered the room through the door, it could then be seen rising from the chimney. No air movement was perceptible in the room, however. The residents admitted that the chimney was too narrow.

Site 3 also is located on the north side and faces south. It has electricity, as do all the other rooms of the dwelling. This room is used as a bedroom and for dining. The walls are covered with paper to the height of 1.5 meters, above which level the loess is rough-cut and not finished with mortar. There is no chimney in this site. Site 4 is used as a storage area and is similar in structure to Site 3. The front wall is brick. Site 5 is in the patio; Site 6 is entirely above ground and outside the complex. The nine rooms of this dwelling are occupied by eight people in four families.

THERMAL PERFORMANCE

Summer temperatures in the Li Songbin dwelling are quite typical of many pit cave dwellings (fig. 129). Temperatures were measured at six sites in this complex, four of which are shown in the graphs. The room facing south (Site 2) is the larger room on that side and its dry- and wet-bulb temperatures have two periods of stability, one during the night and the other during the afternoon. The time before noon shows a gradual increase in temperature. Note that the dry- and wet-bulb temperatures are almost identical during the twenty-four hours because of the high humidity in the air. The difference in temperature between night and day is only 2 degrees C.

The room facing west (Site 4) is similar in temperature pattern. Here too, it is stable during the night and through the morning until noon when it

starts rising. At 4:00 P.M. it becomes uncomfortable because of the high humidity in the air.

The site most difficult with regard to thermal performance is the courtyard (Site 5). Night wet- and dry-bulb temperatures are identical and almost stable. However, the differentiation increases throughout the day and each reaches a peak at 1:00 P.M. The dry-bulb temperature rises to 38 C, much higher than that of the outdoors (Site 6) where the wet-bulb temperature rises to around 28 C in the afternoon. This increase occurs despite some design elements that were expected to ease the temperature of the patio in the afternoon hours, such as its shape (the length being double the width) and the fact that the patio is level with the public alley outside. In spite of this, air circulation was minimal in the courtyard and heat accumulation was intense.

Outdoor temperatures in the open (Site 6) fluctuated at night and increased steadily throughout the day, reaching 34 C at 4:00 P.M. The temperature of the patio had peaked three hours earlier and the differentiation between them was 4 degrees C. On the other hand, at 4:00 P.M. in the patio, the temperature was lower by 3.5 degrees than that of the outdoors. At night, the temperatures of all sites fell within a narrow range, between 23 C minimum and 26 C maximum, bordering on the comfort zone.

Winter temperature measurements display a different pattern from that of summer (fig. 130). The bedroom facing south (Site 2) gains heat from the sun. Its dry-bulb temperature is between 12 C in the afternoon and 9 C at night. Possibly it would require some heating during the night. The difference between wet- and dry-bulb temperatures is constant (2.5 degrees) most of the time. The storage room facing west (Site 4) has a similar pattern but it is cooler, with its dry-bulb temperature registering around 3 C. The temperature of Site 4 fluctuates less than that of Site 2 primarily because of its orientation to the sun; the former receives solar radiation in the afternoon hours only. The dry-bulb temperature of Site 1 (bedroom facing south) is stable with only slight fluctuations: 21 C in the evening and 20 C in the late afternoon. In any case, Site 1 is cooler in summer by 4 degrees than Site 2. Since both rooms have the same orientation, we credit this to the ventilation shaft in Site 1. In the winter, dry-bulb temperature in Site 1 is almost 9 degrees cooler than Site 2. Thus, the advantage of the air shaft is seen in the summer but it detracts from comfort in the winter. The patio (Site 5) in both summer and winter is the most difficult site among the six measured. The patio's dry-bulb tem-

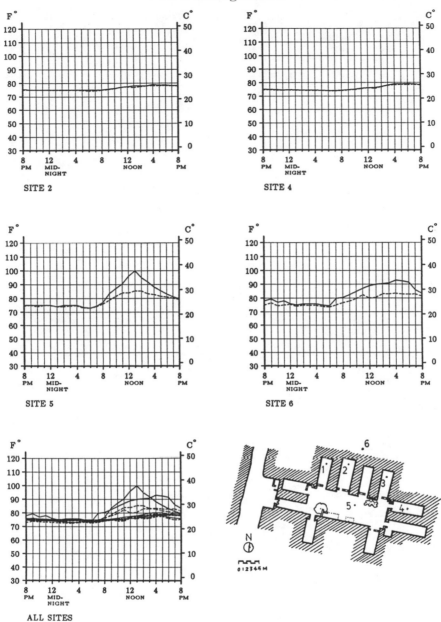

Fig. 129. Summer dry-bulb (solid line) and wet-bulb (broken line) diurnal temperatures, 9–10 August 1984, of the Li Song-bin family dwelling (No. 11).

perature drops below zero (−4.6 C) late at night because of acute air stagnation. The peak, occurring in the afternoon between 2:00 and 5:00 P.M., does not exceed 1.6 C. The minimum temperature at night and the maximum during the day fall below those of the outdoor open space (Site 6). These findings concerning the discomfort in the patio, both in summer and in winter, are contrary to commonly-held opinions. The range of temperature differentiation of all sites, 19 degrees C, is quite wide, from −7 C (wet-bulb) to 12 C (dry-bulb).

The summer relative humidity in Sites 2 and 4 is similar and enormously uncomfortable, reaching almost 100 percent throughout the twenty-four-hour period (fig. 131). Most of the dwellings researched in the other provinces did not reach this high, almost stable, level of diurnal relative humidity. Relative humidity in the courtyard (Site 5) and at the outdoor site shows a different pattern, with the former very high throughout the night (over 95 percent), dropping to a low of 58 percent at 1:00 P.M. because of the heat. Site 6 is also high at night

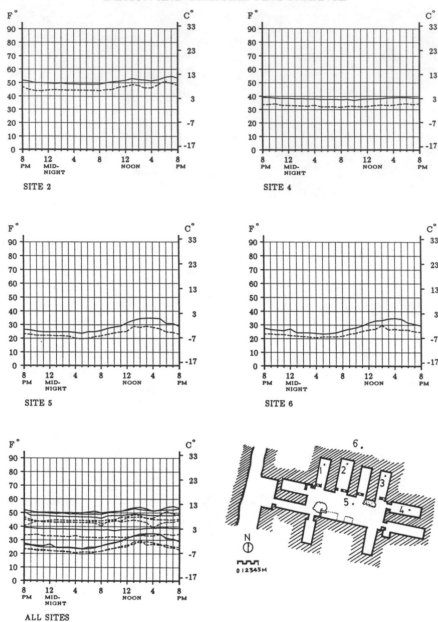

Fig. 130. Winter dry-bulb (solid line) and wet-bulb (broken line) diurnal temperatures, 26–27 December 1984 of the Li Songbin family dwelling (No. 11).

(between 87 and 98 percent), but it is not stable and also drops at 1:00 P.M. to 68 percent.

The winter relative humidity is much lower than in the summer at all sites. Its diurnal fluctuation ranges between 40 and 70 percent. The high is usually at night when air is stagnant and the low point is in the afternoon when temperature rises. In any case, Sites 2 and 4 represent more stable diurnal patterns than those of the patio and the outdoors.

Dwelling No. 12: Liu Xueshi Family Home, Henan Province

ENVIRONS

According to the local people, cave dwellings have been built in Henan for several thousand years, and those of the Luoyang region are the best examples in the province. Luoyang City itself was the capital city of China for a long period (fig. 132).

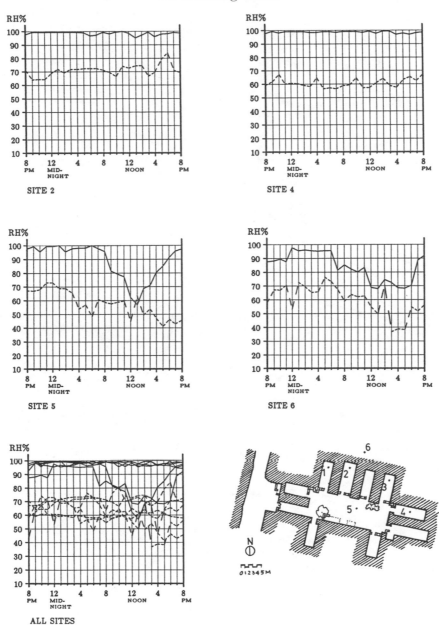

Fig. 131. Summer (solid line) and winter (broken line) relative humidity of the Li Songbin dwelling (No. 11).

The local society of architects estimates that there are 300,000 to 500,000 families living in cave dwellings and earth-sheltered habitats in the mountains of the Luoyang region. If one considers that family size is three to four people in this area, we can estimate that around 1.5 million people live below ground, and among them half a million in cave dwellings.

The region itself is full of historical sites. Among the most interesting are the Longmen Buddhist Caves. These caves were constructed in A.D. 494 and continued in use until the seventh century. The 1,300 caves, with their 10,000 carved images, are located some 14 kilometers south of Luoyang, stretching along the west bank of the Lou He River. It is an ideal site for carving, and the image of Buddha is presented in different sizes and in a progression of styles in the frameworks of cave openings. Another famous site is the old White House Temple, about 12 kilometers east of Luoyang and still in use. The temple is composed of a series of independent buildings arranged linearly, and

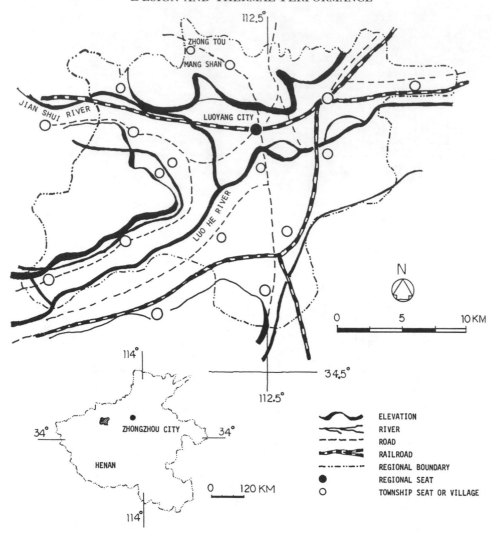

Fig. 132. Mang Shan Township in Louyang City region and Zhong Tou Village, where Dwelling No. 12 of the Liu Xueshi family is located.

many contain sculptures of Buddha and his followers.

LIU XUESHI DWELLING

The agricultural village of Zhong Tou, where we conducted our research, consists of 3,200 people or about 400 families, half of whom live in cave dwellings. The village is located on the high plains and the site itself is flat. The soil is considered to be hard loess. For the last twenty years, new cave dwellings have not been constructed in the village. Moreover, twenty or thirty families abandon their caves every year because of the rain and subsequent high moisture content. The local government has recently shown a willingness to help improve the design of below-ground dwellings.

The cliff cave dwellings of Zhong Tou village feature rooms built with vaulted bricks attached to manmade cliffs and then covered with soil, resulting in earth-sheltered habitats.

The below-ground housing in the village, however, is made up primarily of pit cave dwellings which are arranged in an orderly manner (fig. 133). Most of them are well organized and well built, some constructed a few hundred years ago. The surrounding ground is flat and usually kept clean and clear. At the openings, some of the inhabitants have built overhangs of Mediterranean-style roof tiles to prevent rain from striking the pit walls. In other cases, a low wall of four or five rows of bricks is constructed around the square or rectangular opening. The pit walls are usually cut straight. Depth of the patios is usually 10 to 12 meters below

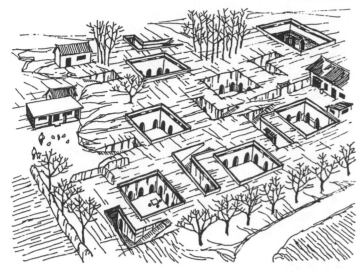

Fig. 133. View of section of Zhong Tou Village, Henan province. The researched dwelling (No. 12) is at the lowest part of the complex.

Fig. 134. Entrance to pit cave dwelling, Zhong Tou Village.

ground level. At a depth of 3 or 4 meters, there sometimes appears a thin, hard, but not completely solid, horizontal layer of loess, about 10 centimeters thick, which becomes the ceiling of the caves and helps to support them. In many cases trees have been planted in the patios, and most have risen above the pit openings. The entrance to the pit patios is by a stairway leading downward with several right-angled turns (fig. 134). A few cave dwellers have constructed a second story for storage space. This is located above and between two first-floor units, thus easing the load on the lower floor. However, the soil left above the second story is hardly one meter thick and we assume that the second floor space would have a high moisture content.

Liu Xueshi, a village school teacher, is head of the family residing in the pit cave dwelling complex analyzed. The dwelling was built about twenty years ago and the overall design is typical of Zhong Tou Village (fig. 135). The entranceway starts its descent from the east and makes two right-angled turns, passes through two arches, and enters the patio from the south. It is composed of paved steps for pedestrians and a ramp to accommodate vehicles and livestock. The last 2 or 2.5 meters is below-ground. The staircase walls are artistic with a combination of brick and locally available yellowish-white stone. It is evident that the villagers have given much attention to the design of the main entrances to their living space and Liu Xueshi's exemplifies this. In accordance with the Chinese custom of preserving privacy, the view into the patio is blocked by a wall.

The patio is rectangular: 9.5 by 7 meters deep (fig. 136). The walls of the courtyard are brick for .75 meters and above that are mortared with a

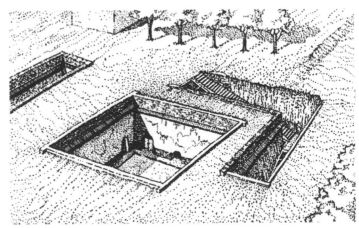

Fig. 135. Bird's-eye view of Liu Xueshi family dwelling (No. 12) in Zhong Tou Village, near Luoyang City, Henan province.

mixture of straw and loess. There is a paved brick walkway around the patio and several trees growing there. The thickness of the soil above the cave rooms is 4.5 meters, which is above the average of other provinces.

Fig. 136. Looking north into the patio of Liu Xueshi family dwelling (No. 12).

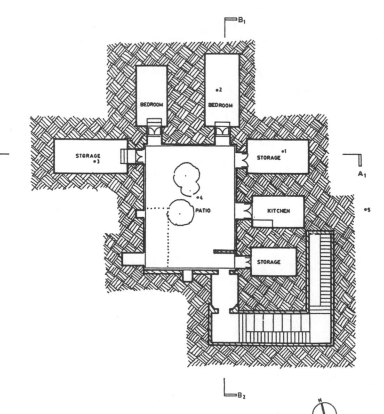

Fig. 137. Plan of Liu Xueshi family dwelling (No. 12).

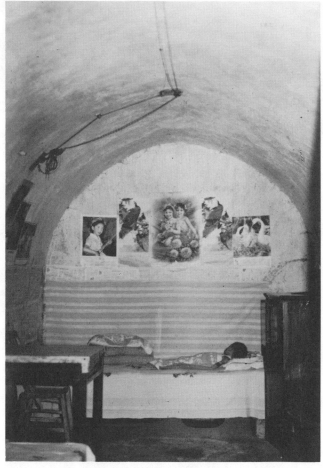

Fig. 138. Interior view of a cave room of the Liu Xueshi family dwelling (No. 12).

Three cave room units are on the east side of the courtyard, with two more on the north and a storage area on the west. The main entrance is on the south, the least desirable location to live (fig. 137). The two north rooms receive the most sun and are used as bedrooms, while all of the other rooms are used either for cooking or for storage (fig. 138). Some cave units are two steps below the level of the patio floor (fig. 139).

THERMAL PERFORMANCE

Temperatures were measured at five sites in the Liu Xueshi family pit cave dwelling complex, four of which are presented here: two rooms, the patio,

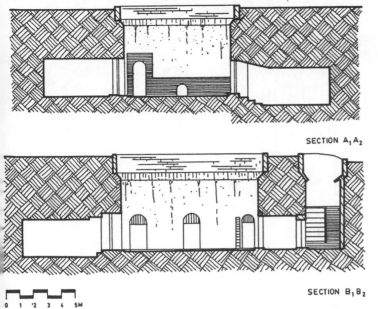

SECTION A₁A₂

SECTION B₁B₂

Fig. 139. Cross sections of Liu Xueshi family dwelling (No. 12).

and the outdoors. Indoor summer temperatures of the two bedrooms (Sites 1 and 2) are generally similar to one another, but differences occur because of their orientations (fig. 140). Site 1, which faces northwest, receives little sunshine, while Site 2, oriented toward the southwest, is influenced during the day by the sun. In both sites there is minimal differentiation between the dry- and wet-bulb temperatures and between the temperatures of the peak hours of the afternoon. The patio (Site 4) shows a major difference between its two temperatures, especially from 11:00 A.M. through the afternoon, and the dry-bulb temperature has two peaks during the day. The patio is warmer in the

Fig. 140. Summer dry-bulb (solid line) and wet-bulb (broken line) diurnal temperatures, 12–13 August 1984, of the Liu Xueshi family dwelling (No. 12).

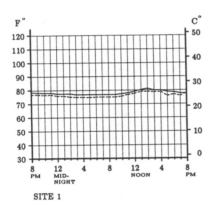

SITE 1

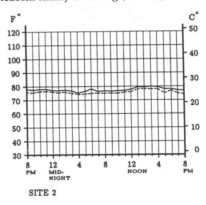

SITE 2

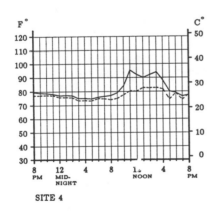

SITE 4

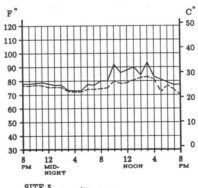

SITE 5

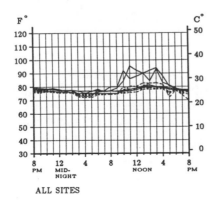

ALL SITES

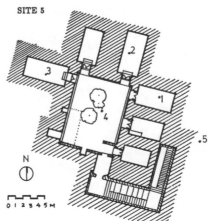

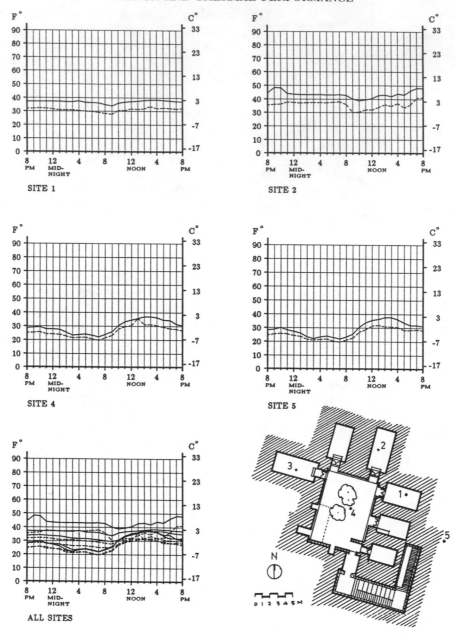

Fig. 141. Winter dry-bulb (solid line) and wet-bulb (broken line) diurnal temperatures, 30–31 December 1984, of the Liu Xueshi family dwelling (No. 12).

afternoon than the outdoors (Site 5) where the temperature rises sharply early in the morning and has three peaks during the day. The range of temperatures of all sites (both wet- and dry-bulb) is between 22 and 27 C, with the exception of the afternoon dry-bulb temperatures in the patio and the outdoors.

Winter temperatures display a different pattern in almost all measured sites (fig. 141). Site 1 temperatures are quite parallel diurnally, with a difference of around 4 degrees. The dry-bulb

temperature is between 0 and 3 C and the bedroom would require some heating to bring it up to a comfortable level. The lowest temperature is at 9:00 A.M. because of the stagnant air that remains overnight in the courtyard. Site 2 temperatures fluctuate throughout the day and the early evening because of the room's orientation to the sun. Dry- and wet-bulb temperatures of the patio and of the outdoors (Sites 4 and 5) are similar to each other diurnally. Temperatures fall steadily throughout the second part of the night and rise at 3:00 A.M. and

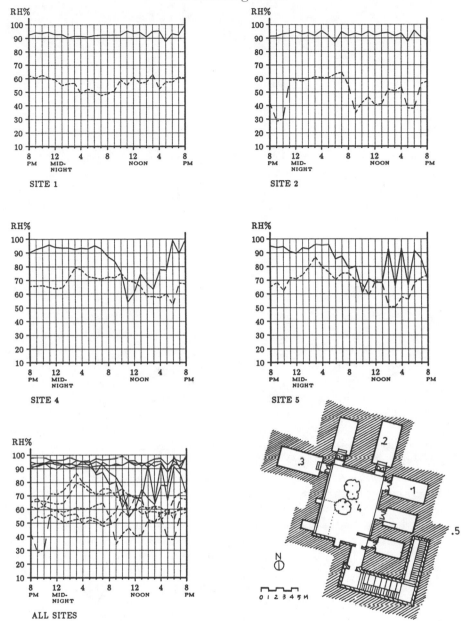

Fig. 142. Summer (solid line) and winter (broken line) relative humidity of the Liu Xueshi family dwelling (No. 12).

again at 7:00 A.M. The rise continues throughout the day to reach a peak level of 3 C from 2:00 to 3:00 P.M., before dropping to zero in the evening. However, it should be noted that this was a very cold day at Zhong Tou Village in Henan.

Calculations of summer and winter relative humidity show a considerable difference between the outdoors and the patio on the one hand, and the two indoor rooms on the other hand (fig. 142). Sites 1 and 2 show high relative humidity in the summer (above 90 percent), while in the winter, Site 1 ranges between 47 and 63 percent, and Site 2

(which fluctuates very much, especially during the afternoon and evening) ranges between 64 percent at 7:00 A.M. and 28 percent at 9:00 P.M. Summer relative humidity is above 90 percent at night in the courtyard and drops sharply to 55 percent at 11:00 A.M. The winter relative humidity of the courtyard also fluctuates, especially at night (between 64 and 80 percent), and drops sharply at 5:00 P.M. to 53 percent. Yet the relative humidity is higher between 10:00 A.M. and noon than that of the summertime. Similar phenomena occur at the outdoor site as well.

4

CONCLUSION

Thermal Performance

Although it becomes clear that the overall thermal performance of the cave dwellings is very beneficial to the inhabitants, there is still a need for design improvement to further minimize discomfort. The contrast in thermal performance between the indoors and outdoors is great both in summer and winter. Any solution should consider the dominancy of summer and winter. While both seasons introduce thermal problems within the cave dwellings, the summer problems are more acute. The winter courtyard temperatures induce air stagnation, while the summer temperatures induce high heat accumulation combined with high relative humidity, especially in the patios of the southeastern province of Henan. Finally, the outdoor ambient temperature differentiation between winter and summer ranges from −18 C in winter to 30 C in summer.

Dwelling units facing south enjoy sunshine, especially in winter, and are preferred for living space. Those units facing west or east have second preference. In cliff type habitats, the dwelling unit should be on a site facing south, east, or west. Problems arise when a pit type dwelling is exposed to less light and sunshine and its north side is dark and damp.

The twelve researched cave dwellings do not require any special cooling systems in the summer. Their thermal performance, however, still has deficiencies in terms of ambient comfort resulting from the combination of heat and high humidity that creates intolerable conditions within the caves and the courtyards.

To reduce the humidity there is a need to establish air circulation by constructing a chimney with a wind catcher high above the cave dwelling that would lead into the lower part of the room unit. Air movement throughout the day will decrease the relative humidity within the cave and thereby improve ambient comfort.

The courtyard's thermal performance constitutes the most serious problem within the cave dwelling complex, especially the pit cave dwelling. It was noticeable that in very cold weather the courtyard was most uncomfortable and that cold air tended to stagnate there late at night. With the construction of the wind catcher and chimney, however, it would be necessary to eliminate penetration of the cold winter air into the room units by closing the doors and windows. On winter nights, although the dwellers' combined human heat radiation would be retained, some additional heating system would still be required to bring temperatures to the ambient comfort range of 22 to 24 C.

In the summer all the pit dwelling courtyards were overheated in afternoon hours and overcooled in the second half of the night. In most cases the afternoon temperatures within the courtyard were much higher than the temperatures in the open area outside the complex. The lack of circulation, and in some cases the evapotranspiration produced by the dense vegetation, made the area intolerable in the afternoon hours. It is our conclusion that the introduction of a passive ventilation system would ease this courtyard condition. A limited amount of vegetation—which should be selected especially for courtyard needs—could be cultivated, however. The plants chosen should not have large leaves, such as banana, which intensify evapotranspiration. Ideal choices are trees with tall trunks free of leaves at the lower and middle levels, yet with an umbrella-shaped leaf mass at the top level. These would provide shade yet still allow air movement within the courtyard. The observed conditions of the courtyards support the recommendation that cliff cave dwellings are more thermally preferable than pit cave dwellings.

Relative Humidity and Ventilation

The most acute problem in the Chinese cave dwellings is lack of air circulation or ventilation. Different from cave dwelling communities in many other places in the world, such as those of southern Tunisia or of central Turkey, the Chinese case is unique because of the combination of rainy and hot conditions in the same season.

The most effective and affordable solution is the introduction of passive air circulation into the enclosed dwelling units. The proposed ventilation system would take air in through a chimney to be constructed above the courtyard level. The warm air coming into the above-ground opening of the chimney and directed by self-rotating wind catchers will be cooled through the long, dark passage, both factors forcing it to flow downward. Also, due to the heated air rising in the courtyard, air will be sucked out of the cave rooms through the windows, and be replaced by new air sucked down the chimney, thus accelerating the rise of the hot air in the courtyard and establishing effective air circulation within the dwelling. Air circulation will cause ventilation and therefore evaporation and reduction of the relative humidity within the dwellings.

Our ventilation experiment near Taiyuan City, Shanxi province, proved that a reduction in humidity is possible when the cave dwelling design is improved.[10] This research project was concerned with solving one problem only—that of the high relative humidity within the dwellings and the courtyard. It is our conviction that this problem resulted from lack of ventilation and air circulation in the cave rooms and in the courtyard. Thus, increase of ventilation will decrease humidity and therefore improve ambient air temperature and comfort.

The site selected for experimentation is an existing three-room cave dwelling of the Wang Xiang Suo family, located in Dian Po Village, Mazhuang County, a few kilometers west of the city of Taiyuan. The room units face north, thus they lack light and

sunshine (fig. 143). The entrance to the three units is by way of the central room and then through passages to the other two rooms. The thickness of the ceiling/roof is more than 7 meters of earth, the total height from the floor of a unit to the surface of the earth is approximately 11 meters, and the three rooms are almost equal in width (2.65 to 2.25 meters). The middle room is used for cooking and storage, while the east and west rooms are used for living and sleeping. Both the east and the west rooms have wide windows facing the terraced courtyard, which is enclosed by a wall on three sides.

The west room, which is longer than the east room, was selected for the construction of the ventilation system. A long chimney constructed entirely of bricks was built at the inner end of the room leading from the floor level up to the surface of the ground above the rooms (fig. 143, A–A1 section). The total length of the chimney is 16 meters, 5 of them above ground, and the interior of the chimney is 1 meter in diameter. At the upper end of the chimney, a wind catcher 1 meter high was made with cloth stretched across a half dome form made of iron wires. The wind catcher revolves as the wind changes direction, and forces the air down into the chimney (fig. 143). At the bottom of the shaft a lightweight wooden divided door was constructed to control wind velocity. The facade window in the west room was open at all times to support air movement into the room from the chimney. The east room was left without changes, its window was closed during the time of the research and there was no air circulation.

Six measurement sites were selected for purposes of comparing temperatures. Sites 1–4 were used for dry- and wet-bulb ambient temperatures (fig. 143). Site 5 represents the dry-bulb temperature measurements of the wall in the east room and Site 6 the wall in the west room. The wind was measured in meters per second outdoors near the opening of the chimney and at the lower part of it within the west room itself. The wind was not measured in the east room since, obviously, it was at zero condition. The research was conducted on 25–26 June 1985, a few weeks after the construction of the chimney, with all measurements recorded on the hour for a period of twenty-four hours. During this twenty-four-hour period the prevailing winds were primarily from the southwest and west.

The indoor and the outdoor wind graphs are compatible (fig. 144). Thus, the air flowing into the chimney at the top and then down through the room created a satisfactory circulation of air in spite of the fact that the wind velocity was not always

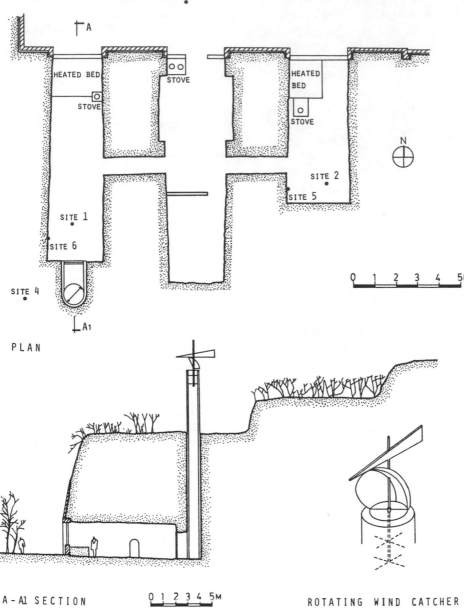

PLAN

A - A1 SECTION

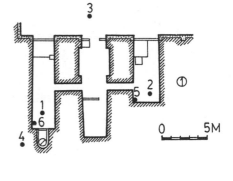

ROTATING WIND CATCHER

Fig. 143. Plan and cross section of Wang Xiang Suo family cave dwelling in Dian Po Village, Mazhuang County, near Taiyuan City, Shanxi province, where experimentation and research was conducted. Note the rotating wind catcher.

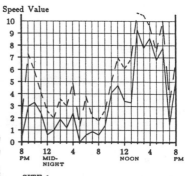

Fig. 144. Diurnal wind speed value outdoors at the top of the chimney near Site 4, (broken line), and at the bottom in the west room (solid line), Wang Xiang Suo family cave dwelling, Dian Po Village, Mazhuang County, Taiyuan City, Shanxi province, 25–26 June 1985.

strong. We believe that the dimensions of the chimney in all its facets are adequate and proved effective. The wind was always sucked down the chimney at the top and drawn into the room. One of the reasons for this wind direction, and not the reverse, is that the heated air rises in the courtyard on which the west room window opens, thus sucking air from the west room to the courtyard and consequently creating air circulation from the chimney into the room. Moreover, when the outdoor warm air enters the top of the chimney, the process of evaporative cooling of temperature and reduction of humidity begins and continues through the long shaded passage into the ventilated room. Even when the courtyard air was cooler, such as during the late hours of the night, circulation continued in the same direction. The main purpose for constructing the chimney was to create air circulation and this goal was achieved. Obviously this circulation decreased relative humidity within the room as well.

During the twenty-four-hour period measured, the peak winds occurred first in the afternoon, and second between 9:00 and 10:00 P.M. We believe that circulation in the first instance was supported by warm air ascending from the courtyard. It was only between midnight and 8:00 A.M. during periods of air stability and low outdoor temperatures that there was low or zero circulation within the west room (fig. 144).

The highest outdoor wind velocity occurred at 2:00 P.M. and obviously was higher than the indoor one. The difference between the two locations is relatively small. In any case, the indoor space enjoyed effective wind circulation.

The ultimate purpose of measuring the temperature was to monitor the humidity in the west room (Site 1) where the chimney was constructed, and determine whether it was reduced in comparison with that of the east room (Site 2) which had no chimney. If reduction occurred, the ambient temperature would be reduced as well, the result of cooling by evaporation through air circulation. As usual, dry-bulb temperatures were higher than wet-bulb temperatures at both sites. Dry- and wet-bulb temperature patterns in Sites 1 and 2 were noticeably similar with differentiation of 5 degrees C. Site 1, the ventilated room, had an almost stable temperature during most of the twenty-four-hour period, according to both wet- and dry-bulb measurements, except for the dry-bulb temperature between 4:00 and 6:00 P.M. Site 1 dry- and wet-bulb temperature graphs show more fluctuation than those of Site 2 during the period measured

(fig. 145). In any case, the relatively low wet-bulb temperature of Site 1 is an indication of more evaporative cooling and therefore of less humidity than in Site 2.

Temperature patterns in the courtyard (Site 3) and at the outdoor location (Site 4) are different in that both fluctuate very much throughout the day with a wide difference between the lowest temperature (in the morning) and the highest temperature (in the afternoon). It is noticeable that the afternoon dry-bulb temperature reached 37.3 C in the courtyard while the indoor dry-bulb temperature of the two rooms was only 24.8 C. In general, the courtyard temperatures were intolerable in the afternoon. The findings were quite different in the morning, especially between 4:00 and 6:00 A.M. at the outdoor (Site 4), and between 8:00 and 10:00 A.M. in the courtyard (Site 3), where the lowest dry- and wet-bulb temperatures occurred due to the fact that both sites are shaded in the morning. The courtyard itself faces north, while the outdoor site is in the shadow of a nearby cliff. In neither case were the temperatures lower than in the two rooms.

Wall temperatures were measured at the ventilated west room (Site 6) and at the unventilated east room (Site 5). The wall temperature shown in fig. 145 is the dry-bulb one. The findings point out that the wall temperatures in the ventilated room (Site 6) were slightly lower throughout most of the twenty-four-hour period than those of the unventilated room (Site 5), with the exception of 11:00 A.M. The cooler temperature in the ventilated room coincides with the air temperature findings and is indicative of lower humidity there.

The construction of the chimney proved to be effective in reducing the relative humidity, especially at the critical time. When we compare the ventilated room (Site 1) with the unventilated room (Site 2), we notice that relative humidity is lower in Site 1 in the hours between 2:00 and 6:00 P.M. (fig. 146), when wind velocity inside Site 1 was at its peak for the twenty-four-hour period. The relative humidity was not lower when wind velocity was almost nil in Site 1.

Basically, without considering the effect of the chimney, the relative humidity within the cave rooms was always higher than that of the patio and the outdoor site due to the high condensation of the moisture inside. The chimney was intended to reduce this high condensation rather than eliminate it. Yet Site 1 itself is not an entirely independent unit as it is connected by corridors to the adjacent two non-ventilated rooms. Although a reduction of

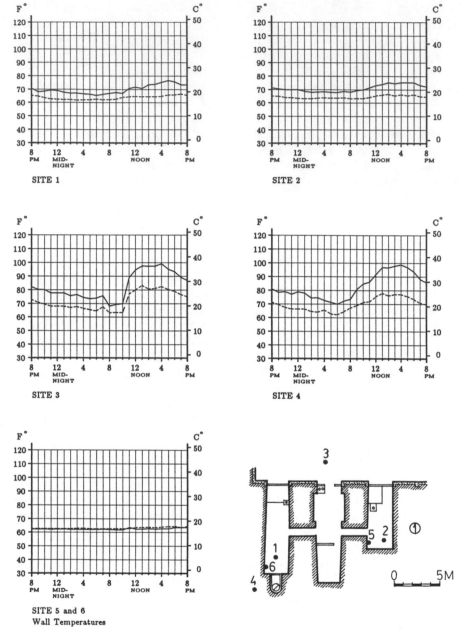

Fig. 145. Dry-bulb (solid line) and wet-bulb (broken line) temperatures of the Wang Xiang Sou family dwelling (Sites 1, 2, 3, and 4). Note the temperature fluctuation among the various sites and the wall temperatures of east (broken line, Site 5) and west (solid line, Site 6) rooms.

relative humidity was attained in Site 1, it remained higher than that of the patio (Site 3) and the outdoors (Site 4).

In conclusion, this pilot research experiment demonstrates that the ventilation system, as designed, proved effective in reducing humidity in the ventilated room and therefore in reducing the temperature. It also indicates that the dimension of the system are effective. However, we would like to

see two improvements: the development of a more flexible wind catcher to follow the wind direction more easily, and enlargement of the facade window in the ventilated room (at present 1 by 0.70 meters). We should also keep in mind that the experiment was run in the summer when there was a need to reduce humidity in the air, but it will be necessary to close the chimney shaft outlet in the winter to avoid heat exchange. We believe that the results of

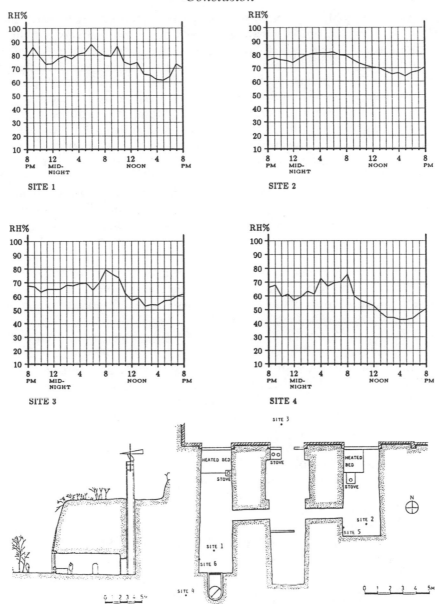

Fig. 146. Summer relative humidity of the four sites measured at the Wang Xiang Sou family dwelling.

this project should be made available to designers of cave dwellings in China. It is especially recommended for the below-ground pit dwellings that support high humidity in the summertime.

Innovative Design Needs

The evolution of the Chinese cave dwelling has taken place over almost four thousand years—before, then along with, the evolution of above-ground housing. Concentrated in a well defined geographical area, they have spread over almost a third of China within the confines of the loess soil zone. Cave dwellings are a synthesis of diversified forces, each of which had different strengths at different time periods. A wealth of experience in design and thermal performance has been accumulated with regard to the Chinese cave dwellings. However, innovations have been relatively minor throughout their long history. Contemporary forces—combining social, economic, climatic and many other factors—are expected to inflict radical changes as the Chinese society undergoes modernization.

It is essential to remember that, different from other countries, China has implemented a strong government policy of limiting urban growth and supporting rural improvement both environmentally and socio-economically. This policy has had a positive impact on rural development. In spite of some reservations among the younger generation about living below ground, the cave-dwelling movement is not diminishing in China. On the contrary, the general trend is toward further growth. With an estimated thirty to forty million people living in some form of subterranean dwelling, urgent consideration should be given to design improvements to meet their needs and to keep in step with the numerous socio-economic changes taking place in the country.

China, more than many other developing countries, has a serious shortage of building materials. Wood has been scarce throughout most of China for many centuries, having been used for fuel or burned to clear land for agriculture. Large-scale industrial, construction, scientific, and urban centers have benefitted from the introduction of advanced technology in the last two decades. However, the introduction of advanced technology, especially mechanization, to the rural areas where housing construction is a pressing issue, is quite slow. In any case, the construction of a cave dwelling unit requires only simple technology that has been available to the farmer for many centuries: the hoe, shovel, and basket are the basic required equipment. The prime building material has been earth in a variety of forms, especially rammed earth, brick, and adobe. Cave dwellings require hardly any extra building materials. They are the "cut and use" form of dwelling. Improvements can be made at the time of construction or later by adding brick facing to the interior vault, walls, and facade; by providing better quality wooden windows and doors; or by adding above-ground units designed in a style similar to that of the below-ground units. The overall deficiency of Chinese cave dwellings is a lack of attention to design details.

The forces that support the cave-dwelling movement in China today are economic and technological. Housing is an acute problem as it is in many developing countries. However, the Chinese have been successful (by their standards) in providing a minimum living space for their people. Almost all Chinese are considered to be employees of the government, and personal income usually borders on the minimum amount required for survival although there has been reasonable salary growth since January 1986. It is worth mentioning that the recent economic reforms in China have taken place at a much greater rate of speed in the rural areas, where the bulk of the cave dwellings are, than in the urban centers. There is also a new liberalization allowing farmers to take advantage of the free enterprise system. It is expected that this income growth policy will continue in the near future. However, it will not be large enough to impede the construction of more below-ground dwellings. Increases in income result in increased consumption, especially of modern furniture and appliances, which in turn establishes a need for a new cave design. The improvement in income will inevitably cause a movement by some of the younger generation out of the cave dwellings to above-ground housing, and the attempt by others to improve the esthetic appearance and the practical usage of their dwellings.

Visual considerations in private rural dwellings have commonly been of low priority throughout most of the world, and the Chinese case is no exception. However, the Chinese cave dwelling compared with the Tunisian cave dwellings, for example, are better designed and articulated, and have considerably more interior design. Basically, Chinese culture is rooted in a significant art and architecture of all types, and the cave dwellings have received a small share of this accomplishment.

PROS AND CONS

Nonetheless, there are some forces that do not support or encourage the development of cave dwellings in China. Throughout the history of China, cave dwellings were, and still are to a great extent, associated with a negative image of poverty and backwardness. It was not customary for rich people to live in cave dwellings before the Communist revolution. This same self image largely exists in China today in spite of the many improvements which have taken place since Mao Tse-tung and his troops constructed large numbers of subterranean dwellings in the Yan'an region during the 1930s and 1940s after the Long March. Generally, young people today want modernization and imitation of the Western lifestyle which is associated with living in an above-ground house. During our field research surveys, we were occasionally told that some young women made it a condition on their future husbands that they would not have to live below ground. If a survey of population age groups residing in cave dwellings today was compared with one of two decades ago, we would anticipate seeing an increase in the cave dwellers' average age.

Among the serious problems that exist in the design of the cave dwellings are shortages of ventilation, light, and sunshine. Other problems are related to demands resulting from improvements in the standard of living, such as the lack of a special space for a kitchen, a living room, shower, and lavatory; the functional division of room usage that can provide some privacy for children and parents; and last, but not least, a design which can accommodate comfortably modern furnishings and appliances—such as a refrigerator, TV, stereo, VCR, sofa, dining tables, closets, and the like. Although there have been some attempts to meet these needs, there should be a breakthrough in design to meet the demand.

It is certainly possible and practical to introduce design innovations that will respond to the new environmental, functional, and social needs of the people. The absence of such a new design may cause an increase in construction of above-ground housing and the consumption of additional good agricultural land for housing. In general, the Chinese home differs from the Western world's style in that it does not designate rooms functionally and absolutely (dining room, kitchen, bedroom, storage, children's, etc.). Contrary to the American style of living, Chinese society has focused basically on pragmatic socio-cultural, economic, and historical evolution; rather than the worship of privacy. Recently, there has been an increasing demand, especially by the young members of the family, for more privacy and for conducting themselves in a so-called modern style, socially and otherwise. This phenomenon is now common in both urban and rural society.

The psychological effect of living below-ground was surprisingly less acute among the Chinese than we had expected, considering the literature describing human reservations about the subject. In our interviews with the Chinese cave dwellers only a few mentioned the psychological aspect as a drawback when we requested their reaction. Apparently it is a matter of acceptance and adjustment since they have lived this way since they were born. However, the feeling of claustrophobia in a confined environment may exist more among the pit cave dwellers than among the cliff cave dwellers. In our opinion, design of the cave dwellings in China, as well as in other regions, should consider two approaches. First, access should be by ascending and not descending. Second, dwellers, when they are indoors, should have direct eye contact with the outdoor environment through a window and be able to feel the rhythms of the natural environment and be a part of it. The Chinese cliff dwellings respond directly to those two design needs.

Internal circulation and arrangement is another weakness in cave dwelling design. In most cases, each room is an independent unit and is connected to the others only through the patio. As it is in the above-ground Chinese house design concept, the patio is the center of privacy and at the same time it is the center of circulation. There are occasional interior corridors connecting the rooms. Except in a few cases however, a room is a dead end with one single entrance from the patio. Thus, in case of hazard, there is a high risk of being trapped inside. The introduction of a corridor running behind all of the rooms and linking them on three or four sides of the cave complex would create a circulation pattern at the rear in addition to the front. The issue of circulation is more acute in the pit cave dwellings than in the cliff dwellings where the rooms of the former surround the deep pit patio which has only one entrance stairway. An emergency stairway can be constructed on the north side of the unit since the other exit is usually located on the south side.

The response of the cliff dwellings to family social needs and to environmental constraints is positive, especially when the dwellings face south. It is our conclusion that the cliff dwelling has greater design advantages over the pit cave dwelling. Most design arrangements of a cave-dwelling complex are dictated by soil constraints, unless cement or other strong building materials are utilized. Most loess soil sites limit the width of one cave dwelling room to about 3 to 3.5 meters at most. The introduction of a cement vault has enabled some below-ground space vaulted units to expand to 5 meters or more. The length, of course, is almost unlimited especially within the cliff dwellings.

Cave dwellings in rural areas have traditionally been an aggregate of disconnected below-ground complexes that formed a village or town. Recently above-ground structures have begun to mushroom within such a village, as the result of the growth of families and the traditional desire of sons or daughters to live near their parents or grandparents. There is also the new tendency of the younger generation to prefer above-ground structures.

Property lines for the cave dwellings are not clear, especially for those located in rural areas. Theoretically, all the houses above ground and the cave dwellings below ground belong to the state. There is, at least in the pit cave dwellings where land is flat, a theoretical line between one complex and another beyond which the digging of a room will not be extended. The space left between two adjacent complexes may range from one to three meters. Reallocation of people within the village is rare. The tenant pays the local government a nominal fee, which may be equivalent to one or two dollars, for the monthly rent of a cave dwelling. A newly married couple may obtain land as well as shelter (above or below ground), when they establish their family. In the mountainous areas, a theoretical border line can hardly exist, and of course it is possible to extend a cliff cave room back into a mountain to an unlimited degree. In general, cave dwellings have been traditionally used for one extended family, and as the family grew, the complex was expanded. Recently we have found a few small unrelated family units residing in a complex, sharing a common patio, or partitioning it among them. Moving from or exchanging units is negotiable between the parties involved and the local government.

Dual usage of land for both agriculture and for cave dwellings has been a commonly accepted practice in China and has been one of the motivating factors for the development of cave dwellings. This practice has shown that there are two major deficiencies. First, agriculture, especially with irrigation, above the pit dwelling can cause serious problems of erosion, water penetration, vertical cavities, and even collapse—particularly during the summer rainy seasons. There is much less of a threat in the cliff type cave dwellings because of the massive soil coverage overhead. Most of the pit cave dwellings today do not use the soil above for agriculture but rather for drying foods, for threshing grain, for piling wheat and for storage. Under such circumstances the soil becomes hard and better resists water penetration. Second, our findings prove that the pit cave dwellings consume much more land than the cliff cave dwellings and are therefore not saving agricultural land as was formerly thought (see table 2). On the other hand, cliff cave dwellings do save land for agriculture since they use terrain unsuitable for cultivation.

Gardens and orchards are commonly maintained in many cave patios. Yet in spite of the advantages for food production they create serious discomfort, particularly for occupants of the pit cave dwelling, by contributing to an increase in humidity through evapotranspiration and by limiting air circulation, especially when dense vegetation exists. Such a patio becomes a trap for heat and is very high in humidity during the summer. The best solution retains only trees with tall bare trunks and leaves at the top that would establish shade and do not prevent air movement.

The cost of cave dwelling construction is at least 50 percent lower than that of equivalent above-ground housing in the same region. The cave dwellings require no special building materials because of the nature of the loess soil unless improvements are made by adding bricks, etc. They consume very little or no wood and no water, they do not require skilled labor or sophisticated technology, and they are expandable as the need arises. Since they are less exposed to weather changes, they last longer than above-ground house units. In addition, dual use of the land reduces the land cost for each use by 50 percent.

Some innovative designs may increase the cost of construction by a small fraction while others, such as the use of a cement vault, require a greater investment. However, it is our estimate that the total cost of the cave unit would still be considerably lower than that of an above-ground house and therefore affordable in the rural community.

Considering their life spans, cave dwellings have far greater economic advantages than above ground housing. In addition, cave dwellings are cool in summer and warm in winter, they provide a comfortable ambience, and they save energy. Economically, they have greater advantages and suitability for Chinese villagers than the above-ground housing. The concept basically has great advantages for a modern designer, who can certainly combine traditional practices with modern standards and norms.

Dust and surface permanence are other issues within the cave dwellings that need design attention. The source of most cave dust is the uncovered walls of the cave and the patio, especially when they were built by the "cut and use" method. The lack of air circulation does not help the situation. One of the traditional ways to halt the dust has been mortaring the vault wall with the yellow soil mud mixed with straw. Yet quite often this mortar crumbles and leaves the interior walls in bad shape. A common solution is newspaper pasted over the vault walls to minimize or eliminate dust accumulation. The ideal remedy, of course, would be to cover all the walls with brick, as many cave dwellers are doing today.

Maintenance below ground is less burdensome than in above-ground houses. This is due to the fact that cave dwellings are simply spaces created within the soil that do not have outside walls. The above-ground house, on the other hand, with exposure to climatic changes (wind, temperature, humidity, dust, etc.) necessitates occasional repairs, painting, and the like. The absence of such structural elements as roofs and a large number of windows in the cave dwellings reduces the maintenance needs and related expenses to a minimum. Maintenance is confined to the walls surrounding the patio, to the inside walls of the vault, and to the soil surface above the cave dwellings. Patio walls are subject to climatic effect and to water erosion and therefore necessitate occasional maintenance to avoid deterioration and crumbling. Here too, the best treatment is brick coverage. The above-ground soil needs to be observed during the rainy season to avoid water infiltration and cavities.

Cave dwellers have traditionally used special below-ground spaces for storage of vegetables, fruit, and grain. A food storage well was dug a few meters deep or a well was dug in a niche in a wall and then deep down. These places are reached by a ladder or a rope and can store fruit and vegetables for several months without decay. The stable low temperature and stable humidity contribute to preserving freshness. Grain is usually stored in one of the cave rooms in huge jars or sacks or piled loose on the floor. In this case too, stability of temperature and humidity help preservation. The designer should study and consider this experience when planning a modern below-ground habitat.

In Chinese society in general, and in rural communities in particular, rooms are not designated for a single type of usage as is regularly done in the West. Cave dwelling room functions and usage are mixed—living, cooking, sleeping, storage, and so on. Cooking, especially in the summer, takes place in the patio or outdoors in the open. Cooking facilities do exist within the cave units however, and in winter cooking is mostly done in the bedrooms to add to the heat required for comfort. Recently there has been a tendency to have rooms designated for single or dual functions.

Odors and smells exist in the cave dwellings in different degrees, more so in the pit dwellings than in the cliff dwellings, due to the lack of ventilation, the result of cooking, livestock, the open sewage network, and the proximity of the open toilet. Livestock within the pit cave dwellings constitute somewhat more of a health hazard for the residents than livestock in the cliff cave dwellings. Adequate bath and toilet facilities are items that have never had a high priority with the village cave dwellers. Arrangements for the toilet are mostly located near the entrance to the complex and outside the cave dwelling itself, as shown in some of the twelve case studies. There is simply a space above ground enclosed by three and one-half mud walls where the residents are supposed to cover their waste with earth using a shovel left nearby. The waste then is used by the farmer for fertilizing the cultivated land. No running water is available and conditions are extremely unhealthy. Odors may pass into the complex. The surroundings may be muddy, especially in the rainy season, which makes conditions worse. The bath, on the other hand, is treated differently from that in the West. Males use the river if there is one nearby. In other cases, mostly common with the females, they clean themselves with a wet cloth using a basin of water. Traditionally soap was seldom available but is more common now. Laundry is usually done at a nearby river or stream by the age-old manner of washing and beating the clothes and hanging them out to dry. Cave dwelling designers should consider including an indoor shower; however, a toilet would be more acceptable to the dwellers if it were located outside the structure. The combination of the two might be acceptable if modern facilities and running water with adequate drainage were provided. In any case, bath and toilet inadequacy are major issues for designer consideration, along with a new sewage system. A seepage system that would allow waste to be used for agriculture is one suggested solution.

Seismic impact can be a serious problem for the Chinese cave dwellers. Basically northern China and the five provinces where the cave dwellings are concentrated is a region subject to earthquakes. Tectonic faults intersect the region especially in the south and east. Physical security of the residents is threatened by intensive erosion, collapse, and by seismic movement. Loess soil is very efficient and holds itself firmly when dry. However, it has a tendency toward vertical erosion, with resultant cavities that can cause collapse in the rainy season. Loess soil varies from one place to another and so does the quantity and frequency of precipitation. Therefore, norms and standards in design should be made part of the building code for each region. Seismic conditions are much more serious problems, and here too, a specially designed code should be enforced in each individual region.

The facade of the cave dwelling unit is subject to climatic changes over time. It may weaken and collapse under ordinary conditions and especially under seismic stress. Such a collapse may block the entrance and trap people inside. Many Chinese use a covering of bricks with ten to fifteen degree backward inclination of the facade that has proven to be effective in most cases. Niches in the interior walls should be avoided as they weaken the structure. The pointed arch is most common and minimizes seismic effect. However, the cave dwelling designer must consider this sensitivity to earthquakes and provide an alternate escape route in case of collapse. Given the earthquake and erosion aspects, the Chinese experience shows that the cave dwelling ceilings should be at least 3.5 meters below the soil surface in order to ensure stability.

Chinese studies and our own research show that cave dwellings are safer than above-ground houses made of stone or brick during earthquakes, and that basically the cliff cave dwellings are much better off than the pit cave dwellings. The latter are more subject to cracks and collapse.

The danger that someone will fall into the pits of a below-ground village exists but, practically speaking, seldom occurs. Most of the pits are not fenced and yet the people, even the children, are used to them and avoid accidents. There are virtually no wild animals wandering around and the domesticated animals are normally tied or penned. The sight of herds of cattle or sheep grazing over large areas of open pasture is confined to the grasslands of Gansu, Inner Mongolia, and Xinjiang far to the north and northwest.

Fire does not constitute a serious threat to the cave dwellings. The traditional cave rooms have few fixtures and the bed is usually made of mud bricks. The cave dwellers use coal for heating or cooking, and in summer cooking mostly takes place outdoors. The modern cave dwellings may be subject to fire hazards with their modern furniture and beds made of wood. And modern cave dwellings often have wooden facades (from 1 to 3 meters above the ground) to hold the glass windows and to enhance appearance. In most cases there is no water supply to put out a fire. However, the hazard is minimal.

Any design should be planned to avoid water penetration from erosion and the drainage process. The uniform loess soil is subject to extensive vertical erosion, thus creating cavities and funnels. Maximum humidity within the soil should not exceed 15 to 20 percent. Trees and shrubs encourage cavities to begin and should not be planted above, or nearby, the cave dwellings. To avoid water penetration within the soil it would be desirable to install plastic sheets horizontally 30 centimeters below the surface above the cave dwellings. It may also be necessary to grade the surface so that water can drain away quickly and not seep into the soil. Because of their location and great depth below the surface, cliff cave dwellings are more secure than pit dwellings in this regard.

Until very recently, cave dwelling design did not involve architects, with the farmer and the urban cave dwellers left to initiate their own dwelling configurations. Yet there are environmental constraints that have limited the imaginative ability of cave dwellers, such as the loess soil characteristics and attendant technological conditions. Very little research has been done to study those and other constraints to improve cave dwelling design. Moreover, in spite of the Communist revolutionary policy of giving priority to the improvement of the standard of living of farmers and workers, cave dwellers were out of the mainstream of thought of their leaders. Research on cave dwellings and the introduction of relevant and applicable designs for improvement have been undertaken only recently by the Architectural Society of China. A national committee, the Cave Dwelling Investigation Group, was formed by the Architectural Society of China, headed by the prominent architect Ren Zhenying, with the support of five provincial subcommittees. This group has held three national symposia on the subject, and the Architectural Society of China has sponsored an international conference on earth architecture (Beijing, November 1985). The Cave Dwelling Investigation Group has introduced and advanced ideas for cave dwelling improvement and has built experimental models. These design improvements are now being felt in some areas, such as Shanxi and Gansu provinces. A model neighborhood has also been constructed near Lanzhou City, Gansu.

In addition, a team of Japanese academicians and practitioners has researched the Chinese cave dwellings thoroughly and developed comprehensive findings on the subject which will certainly help to improve the dwellings in the future. The product of all these activities of the Chinese and the Japanese seems to be of reasonably good quality. However, there is still a need to learn from the modern experience of earth-sheltered habitat systems of the United States and other developed countries.

EARTH-SHELTERED HABITAT

It is our conclusion that one design concept that will meet most of those above-mentioned requirements for improvements is the earth-sheltered habitat combined with the below-ground Chinese cave-dwelling.[11] This synthesis uses the same principle as the vaulted extendable unit to be built on the open excavation of the terraced cliff and then covered with one-half to one meter of earth to be made level with the land above the dwelling. The unit itself should be built from locally made bricks, with a slanted ceiling toward the rear to maximize the reception of light and sunshine. The brick con-

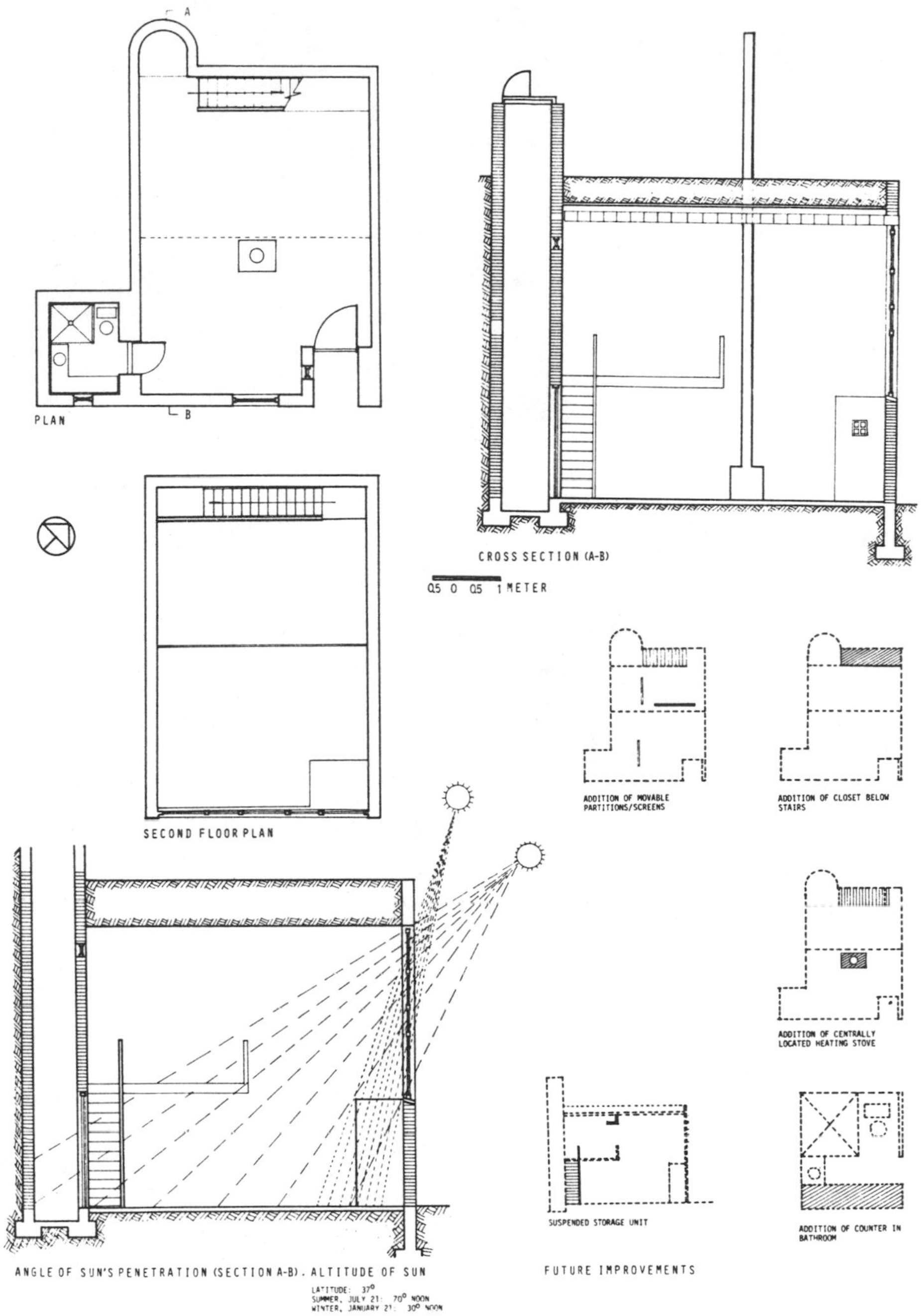

PLAN

SECOND FLOOR PLAN

CROSS SECTION (A-B)

0.5 0 0.5 1 METER

ADDITION OF MOVABLE
PARTITIONS/SCREENS

ADDITION OF CLOSET BELOW
STAIRS

ADDITION OF CENTRALLY
LOCATED HEATING STOVE

ANGLE OF SUN'S PENETRATION (SECTION A-B). ALTITUDE OF SUN

LATITUDE: 37°
SUMMER, JULY 21: 70° NOON
WINTER, JANUARY 21: 30° NOON

SUSPENDED STORAGE UNIT

FUTURE IMPROVEMENTS

ADDITION OF COUNTER IN
BATHROOM

Fig. 147. Plan of earth-sheltered cave dwelling, designed by
the author, near Linfen City, Shanxi province. The dwelling
was constructed by the Architectural Scientific Academy of
Shanxi.

struction of the vault calls for greater width than the
conventional three meters of the historically derived cave design to accommodate modern furnishings, accordingly a new width can be five to six
meters. Similarly, the height at the cave front can
be increased to five or six meters, facilitating the
construction of two floors within the space, thus

doubling the space of the unit at minimum ex-
pense.

The following is an introduction of such an earth-
sheltered dwelling designed by the author with
Chinese collaboration. This dwelling was constructed near the city of Linfen, Shanxi, in the
summer of 1985 and is currently inhabited (figs. 147

A. CROSS SECTION

B. FACADE

C. PERSPECTIVE: COURTYARD (10 X 10 M.) AND OUTDOOR KITCHEN (3 X 4 M.)

D. POSSIBLE ARRANGEMENT OF UNITS; SHARED BATHROOM PLUMBING AND VENTILATION

Fig. 148. More details of the plan shown in Figure 147.

and 148). Research monitoring of this dwelling is currently being conducted by the Architectural Scientific Academy of Shanxi province.

The problems to be solved by this earth-sheltered dwelling design are confined to the most common types existing within the conventional cave dwelling:

1. Limited light penetration, especially in winter.
2. Absence of ventilation within the dwelling units, and/or within the courtyard in the case of the pit type, especially critical in summer.
3. Existence of high relative humidity within the dwelling during the summer rainy season.
4. Existence of humidity in the unit walls and ceilings; intensive infiltration of water can cause collapse.
5. Need to accommodate new furniture for modern living standards in the unit; introduce space for social gatherings of the extended family and friends, and more privacy for each family member.
6. Need to provide low-cost dwellings in the existing tradition of below-ground housing.
7. Need to introduce new elements not considered in the previous standardized design; such as toilet unit, shower unit, closets, and an independent kitchen.

The new design principles being introduced are:

1. Problem solving: to respond to the above-mentioned problems and introduce a design that solves or eases them.
2. Cost: the cost should be affordable, not more expensive than a contemporary brick-covered unit.
3. Earth Sheltered: the unit should be earth sheltered with a maximum soil cover of one meter, yet all the structure is below ground on the terraced cliffside.
4. Orientation: southeast.
5. Integration of the unit within the natural environment; especially responsive to solar needs.

The dwelling plan should be flexible in design to provide alternatives for improvements, for changes, and for additions:

1. To enable the family to improve the dwelling by such factors as the construction of built-in closets, a shower, a bench in the toilet room, or a hanging storage closet at the upper front of the second floor.
2. To develop low, mobile, internal walls made of simple lightweight materials that can be used as dividers to create more privacy for family members with guests; especially for the women and children.
3. To allow construction of some small parts of the dwelling at a later date when the family has settled and improved its financial condition, or deferred so that the dweller can complete construction himself; such as the railing for the stairway, the counter and bath for the toilet room, and the bricks in the upper portion of the innermost wall.
4. Kitchen designed as separate room to be lo-cated in the patio year-round to avoid heat within the dwelling in summer.

Cave dwellings are a traditional way of life that have merit for modern accomplishment. The current Chinese push for modernization and Westernization can negatively affect their continued use. The government is interested in improving the existing cave dwellings and in introducing ideas for a modern below-ground dwelling using the same concepts.

The introduction and implementation of innovative ideas should meet the needs and values of Chinese society. A new design should be economically feasible for the average Chinese family, especially the rural family; consume a minimum amount of building materials; utilize an affordable level of technology; evidence durability; meet the socio-cultural tradition of Chinese house principles (privacy, intimacy, enclosure); accommodate modern household furnishings; introduce much more sunshine and light; provide living space to meet emerging familial and leisure activities; introduce passive ventilation systems; and lastly, still retain positive thermal performance advantages.

The study of Chinese cave dwellings points out the need for much design improvement. Considerations should be given to the social and psychological needs of the cave dwelling residents, and generally to more proper use of below-ground thermal performance advantages. With the increasing interest throughout the world in multiple uses of land by designing below-ground space appropriately, the Chinese experiences can prove most valuable. In conclusion, these cave dwellings are an excellent research laboratory from which we can gain knowledge and experience on earth-sheltered space design and thermal performance.

APPENDIX 1: GLOSSARY

YAODONG	-	窑 洞	Cave dwelling cliff or pit integrated within the soil.
YAO	-	窑	Cave; kiln for burning bricks, limestone, clay products or glass.
KAOSHAN YAO or AI YAO	-	靠山 窑窑 崖	Open-backed cave dwellings-cliff cave dwelling or the mountain.
TIAN YAO	-	天 窑	"Sky cave." Pit type cave dwelling of one or more stories, so-called especially in Henan province.
GAOYAOZI	-	高 窑 子	"High cave." Cave dwelling with two or more floors, so-called in Longdong area.
JIEKOU YAO	-	接 口 窑	Vault-form cliff cave dwelling with above ground, earth-sheltered extension made of bricks or stone.
GU YAO	-	箍 窑	(Gu = vault, Yao = cave). Above-ground earth-sheltered habitat using arch of bricks or stone.
LIANG	-	梁	Ridge.
MAO	-	峁	Hillock.
PING	-	坪	Hill with flat top.
KANG	-	炕	Heated bed (hypothermic bed).

JAO	-	窖	Underground food storage pit or root cellar.
FALA	-	坴垃	Bricks cut from lake mud or river bank deposits.
JIANZHU	-	建筑	Structure, architecture.
DISHANG JIANZHU	-	地上建筑	Above-ground structure.
DIXIA JIANZHU	-	地下建筑	Below-ground structure.
TU JIANZHU	-	土建筑	Earth structure.
YANTU JIANZHU	-	掩土建筑	Earth-sheltered structure.
YUANZI	-	院子	Common term for courtyard.
YUAN	-	元	Chinese basic currency, similar to U.S. dollar.
QIAN YUAN	-	前院	Front yard.
HOUYUAN	-	後院	Back yard.
ZAYUAN	-	杂院	A yard of miscellaneous or mixed functions; a compound occupied by many households.
SANHEYUAN	-	三合院	Traditional above-ground Chinese house with three sides enclosed by buildings.
JING YUAN	-	井院	Small yard.
TIANJING YUAN or	-	天井院	Pit cave dwellings.
DIKENG YUAN		地坑院	
SHENKENG	-	渗坑	Seepage pit in the center of courtyard.
MU	-	畝	Unit of measure of land equal to 0.1647 acres.

APPENDIX 2: CHRONOLOGY OF CHINESE DYNASTIES

Xia	c. 21st century–16th century B.C.
Shang	c. 16th century–11th century B.C.
Zhou	c. 11th century–221 B.C.
Western Zhou	c. 11th century–770 B.C.
Eastern Zhou	770–221 B.C.
Spring and Autumn Period	770–476 B.C.
Warring States Period	475–221 B.C.
Qin	221–207 B.C.
Han	206 B.C.–A.D. 220
Western Han	206 B.C.–A.D. 24
Eastern Han	25–220
Three Kingdoms	220–280
Wei	220–265
Shu	221–263
Wu	222–280
Jin	265–120
Western Jin	265–316
Eastern Jin	317–420
Southern and Northern Dynasties	420–589
Southern Dynasties	420–589
Song	420–479
Qi	479–502
Liang	502–557
Chen	557–589
Northern Dynasties	386–581
Northern Wei	386–534
Eastern Wei	534–550
Western Wei	535–557
Northern Qi	550–577
Northern Zhou	557–581
Sui	581–618
Tang	618–907
Five Dynasties and Ten Kingdoms	907–979
Song	960–1279
Northern Song	960–1127
Southern Song	1127–1279
Liao	916–1125
Kin	1115–1234
Yuan	1271–1368
Ming	1368–1644
Qing	1644–1911

Source: *China Facts and Figures: 4,000-Year History.* (Beijing: Foreign Languages Press, 1982), 7.

APPENDIX 3: QUESTIONS TO THE CAVE DWELLERS

Site Selection

How does he select a site: orientation, proximity to his fields, etc?

What are his criteria?

Does he consult with others if he is inexperienced?

Pre-Design

How does he plan the cave before construction begins?

Does he visit other caves before designing his own?

How does he calculate the scale, dimensions, height, etc?

Construction Process

What are the stages (one by one) of his construction process?

Describe in detail each stage.

What difficulties has he met during construction?

What are the tools that he uses? Make drawings of the tools.

Does he work alone or does someone help him? If so, who?

In which season does he do the digging and why?

How many days will it take him to dig one room 6 by 9 by 5 meters high?

How many hours a day does he dig?

Does he take breaks in the middle and why?

What does he do with the earth that he excavates?

Do other members of the extended family help?

Does he get permission for the construction?

Does he pay costs:

> How much does it cost him to dig one room 5 by 9 by 5 meters high?
>
> What does the cost cover: labor, equipment (tools), construction materials, etc?
>
> Does he get any financial support from outside?

Perception

Does the dweller like to live in a cave dwelling if he is given the choice of an above-ground home, and why?

Does his wife like it?

How old is the cave dwelling?

APPENDIX 4: STATISTICAL TABLES FOR THE DWELLINGS

Diurnal Temperature Measurements

COUNTY: Liquan
PROVINCE: Shaanxi
SETTLEMENT NAME: Gao Jia Team Village
FAMILY/HOUSE NAME: Gao Ke Xi

House No. 1 — Summer
July 11-12, 1984
Degrees in Centigrade

Time	SITES											
	1		2		3		4		5		6	
	Dry	Wet	Dry	Wet	Dry	Wet	Dry	Wet	Dry	Wet	Dry	Wet
8:00 pm	21.7	20.8	21.8	21.1	21.9	21.0	22.4	21.8	25.6	23.1	24.9	22.8
9:00	21.4	20.7	22.1	21.4	21.6	21.2	22.1	21.8	25.2	23.1	24.5	23.3
10:00	21.2	20.6	22.0	21.6	21.4	20.9	22.3	21.8	24.9	23.1	23.9	22.8
11:00	21.2	20.6	21.8	21.3	21.2	21.0	22.0	21.5	24.5	22.2	23.4	22.2
Midnight	21.1	20.6	21.3	21.0	21.1	20.8	22.0	21.6	23.5	22.3	23.3	22.2
1:00 am	21.0	20.4	21.2	21.1	21.2	20.8	21.8	21.4	23.3	22.0	23.4	22.4
2:00	20.6	20.3	21.1	20.8	21.1	20.7	21.8	21.3	23.9	21.4	24.5	22.2
3:00	20.7	20.2	21.2	20.7	20.9	20.6	21.8	21.2	23.4	21.7	23.9	22.0
4:00	20.6	20.1	20.9	20.4	20.7	20.6	21.6	21.2	23.4	21.2	23.3	21.7
5:00	20.6	20.1	21.0	20.6	20.7	20.4	21.4	21.1	22.6	20.9	23.4	21.1
6:00	20.6	19.9	20.7	20.7	20.6	20.3	21.3	21.0	22.2	20.6	23.5	21.3
7:00	20.5	19.8	20.7	20.3	20.6	20.3	21.2	20.8	22.6	20.7	22.4	20.6
8:00	20.2	19.8	20.7	20.2	20.6	20.3	21.1	20.9	22.8	20.9	22.7	20.9
9:00	20.2	19.8	20.6	20.2	20.6	20.1	20.9	20.6	21.7	20.9	21.7	20.6
10:00	20.1	19.8	20.6	20.6	20.6	20.6	21.2	20.9	21.7	20.7	21.7	21.1
11:00	20.1	19.7	20.1	20.3	20.5	20.2	21.1	20.7	21.7	20.6	21.4	20.9
Noon	20.1	19.7	20.6	20.2	20.3	20.1	21.2	21.1	22.3	21.2	23.1	22.0
1:00 pm	20.2	19.8	20.6	20.3	20.4	20.1	21.2	21.1	23.1	21.7	22.8	21.7
2:00	20.3	19.8	20.6	20.3	21.1	20.3	21.2	21.1	24.0	21.7	22.8	21.7
3:00	20.3	19.8	20.6	20.5	20.6	20.3	21.4	21.3	24.0	22.0	22.9	21.7
4:00	20.3	20.1	20.6	20.4	20.6	20.4	21.4	21.1	23.9	21.7	22.8	21.7
5:00	20.3	19.8	20.6	20.3	20.6	20.3	21.3	21.1	24.2	21.7	23.1	21.7
6:00	20.3	20.0	20.6	20.5	20.6	20.3	21.3	21.1	24.5	22.3	24.4	22.2
7:00	20.4	20.1	20.6	20.6	20.6	20.2	21.3	21.1	23.7	21.7	23.4	21.7
8:00	20.4	19.9	20.6	20.4	20.6	20.2	21.3	21.1	23.1	21.7	23.5	21.7

Diurnal Temperature Measurements

COUNTY: Liquan
PROVINCE: Shaanxi
SETTLEMENT NAME: Gao Jai Team Village
FAMILY/HOUSE NAME: Gao Ke Xi

House No. 1 — Winter
Dec. 9-10, 1984
Degrees in Centigrade

Time	SITES											
	1		2		3		4		5		6	
	Dry	Wet	Dry	Wet	Dry	Wet	Dry	Wet	Dry	Wet	Dry	Wet
8:00 pm	9.2	7.2	8.9	7.2	8.4	6.2	8.4	6.4	4.8	3.3	3.3	2.3
9:00	9.4	7.6	9.1	7.3	8.9	6.7	10.1	7.3	4.4	3.1	3.1	2.3
10:00	9.5	7.7	9.1	7.7	9.0	6.7	10.1	7.6	3.4	2.3	3.1	2.0
11:00	9.5	7.7	9.3	7.7	9.1	6.6	10.2	7.6	3.8	2.4	3.1	2.0
Midnight	9.5	7.8	9.4	7.9	9.3	6.4	10.3	7.7	3.4	2.3	3.1	2.0
1:00 am	9.6	7.8	9.4	7.7	9.3	6.6	10.0	7.8	3.5	2.8	2.8	2.0
2:00	9.7	7.8	9.5	7.4	9.3	6.7	10.0	7.8	3.4	3.1	2.8	2.0
3:00	9.7	7.8	9.5	7.7	9.4	7.2	9.9	7.8	3.2	2.4	2.4	1.9
4:00	9.8	7.9	9.6	7.8	9.5	7.4	9.8	7.8	2.8	2.0	2.4	1.9
5:00	9.8	7.9	9.6	7.8	9.5	7.4	9.8	7.8	2.8	2.0	2.8	2.0
6:00	9.9	7.8	9.7	7.7	9.6	7.3	9.8	7.8	3.1	2.0	2.8	2.0
7:00	9.9	7.7	9.8	7.8	9.6	7.0	10.1	7.6	2.7	2.0	3.1	2.2
8:00	9.9	7.5	9.9	7.8	9.6	7.3	10.1	7.6	2.4	2.0	3.3	2.3
9:00	9.1	6.0	8.4	5.6	8.6	5.6	9.1	5.7	3.1	2.4	2.9	2.2
10:00	8.4	6.0	7.8	6.0	8.3	5.6	8.3	5.6	3.3	2.7	2.8	2.3
11:00	8.5	6.0	7.8	5.1	8.1	5.3	7.9	5.2	3.9	2.9	3.8	2.8
Noon	8.4	6.0	7.6	5.1	7.8	5.2	7.8	5.6	4.4	3.7	3.7	2.8
1:00 pm	8.4	5.8	7.6	5.4	8.1	5.6	8.1	5.7	5.1	4.3	4.3	3.4
2:00	8.4	6.2	7.7	5.2	8.4	5.7	8.4	5.6	5.1	4.3	4.4	3.8
3:00	8.5	6.1	7.8	5.9	8.5	5.8	8.4	5.8	5.1	3.7	4.8	3.4
4:00	8.4	6.0	7.8	6.1	8.4	5.6	8.3	5.6	4.3	3.3	4.1	3.1
5:00	8.5	6.2	7.6	5.8	8.3	5.5	7.9	5.7	3.8	3.1	3.8	2.8
6:00	9.0	7.3	8.5	6.8	8.0	5.5	9.2	6.1	3.1	2.3	3.3	2.4
7:00	9.1	7.7	8.9	7.6	7.8	5.7	9.6	5.6	3.1	2.3	2.8	2.3
8:00	9.4	7.8	9.2	7.9	8.9	7.0	10.4	7.3	2.4	2.0	2.8	2.0

Relative Humidity Calculations

COUNTY: Liquan
PROVINCE: Shaanxi
SETTLEMENT NAME: Gao Jai Team Village
FAMILY/HOUSE NAME: Gao Ke Xi

House No. 1   Summer

Time	SITES					
	1	2	3	4	5	6
	RH%	RH%	RH%	RH%	RH%	RH%
8:00 pm	93.2	94.1	93.3	95.8	82.9	85.3
9:00	94.5	95.3	96.6	98.3	85.7	91.2
10:00	95.7	97.0	96.1	96.6	87.1	91.9
11:00	95.7	95.7	98.7	96.6	83.7	90.7
Midnight	95.7	97.8	97.8	97.0	91.1	91.8
1:00 am	96.2	98.7	97.4	96.6	90.3	92.6
2:00	97.8	97.4	96.5	95.7	82.4	83.7
3:00	96.1	96.5	97.8	95.3	87.9	86.2
4:00	95.6	96.5	99.1	96.6	84.1	88.7
5:00	95.6	96.5	98.2	97.0	87.8	83.4
6:00	94.8	99.5	97.8	97.8	88.5	84.2
7:00	94.8	96.9	97.3	96.5	86.6	86.9
8:00	96.9	96.1	97.3	98.7	85.9	87.0
9:00	96.5	96.9	96.1	97.4	93.7	91.6
10:00	97.3	100.0	100.0	97.8	92.0	94.9
11:00	96.5	100.0	97.3	96.5	92.0	95.7
Noon	96.5	96.5	98.2	99.1	91.7	91.8
1:00 pm	96.5	97.8	97.3	99.5	89.8	91.8
2:00	95.6	97.8	94.0	99.1	83.6	91.8
3:00	96.0	99.1	97.8	99.1	85.4	91.4
4:00	97.8	98.2	98.6	97.4	84.3	91.8
5:00	96.0	97.8	97.8	97.8	82.5	89.4
6:00	97.3	99.1	97.8	98.2	84.4	84.4
7:00	96.9	99.5	96.9	98.2	86.1	87.9
8:00	96.0	98.6	96.9	98.2	89.8	87.2

Diurnal Temperature Measurements

COUNTY: Yan'an Region
PROVINCE: Shaanxi
SETTLEMENT NAME: Wang Jia Terrace
FAMILY/HOUSE NAME: Gou Shengzhi

House No. 2   Summer
July17-18, 1984
Degrees in Centigrade

Time	SITES											
	1		2		3		4		5		6	
	Dry	Wet	Dry	Wet	Dry	Wet	Dry	Wet	Dry	Wet	Dry	Wet
8:00 pm	21.2	18.4	21.2	19.3	22.3	20.1	17.7	17.4				
9:00	21.8	19.2	21.2	19.0	22.1	20.2	17.4	17.3				
10:00	21.9	19.0	21.3	19.2	22.3	21.1	17.6	17.4				
11:00	21.8	19.2	21.4	19.0	22.3	21.2	17.4	17.6				
Midnight	21.8	19.4	21.6	19.4	21.9	19.8	17.4	17.3				
1:00 am	21.2	18.8	21.3	18.9	21.8	19.8	17.4	17.3				
2:00	21.1	18.6	21.2	18.9	21.5	19.5	17.3	17.1				
3:00	20.8	18.6	21.2	19.0	21.5	19.4	17.3	17.1				
4:00	20.7	18.5	21.1	19.2	21.4	19.4	17.2	17.1				
5:00	20.6	18.7	21.1	19.4	21.3	19.4	17.3	17.1				
6:00	20.6	18.6	21.2	18.9	21.3	19.4	17.4	17.3				
7:00	20.6	18.5	21.2	18.5	21.3	19.2	17.7	17.3				
8:00	20.2	18.4	20.9	18.4	21.3	19.3	18.2	17.4				
9:00	21.1	18.6	20.9	18.8	21.7	19.6	18.4	18.1				
10:00	21.0	18.8	21.1	18.9	21.4	19.5	19.1	18.4				
11:00	21.2	18.9	21.2	19.0	21.3	19.6	20.0	19.1				
Noon	21.3	19.2	21.2	19.2	21.4	19.7	20.3	19.8				
1:00 pm	22.2	19.8	21.2	19.0	21.7	19.6	21.3	18.8				
2:00	21.9	19.7	21.2	18.7	21.7	19.6	20.7	19.4				
3:00	22.1	19.4	21.2	19.0	21.7	19.9	20.6	19.2				
4:00	21.6	19.4	21.1	18.4	22.3	19.9	20.4	19.0				
5:00	21.2	19.0	20.9	18.4	22.2	19.8	19.8	18.4				
6:00	21.4	19.2	20.9	18.4	22.2	20.1	19.5	18.5				
7:00	20.9	18.7	21.1	18.4	22.0	19.5	18.9	18.2				
8:00	21.2	18.6	21.1	18.4	21.8	19.2	18.7	17.9				

Relative Humidity Calculations

COUNTY: Liquan
PROVINCE: Shaanxi
SETTLEMENT NAME: Gao Jai Team Village
FAMILY/HOUSE NAME: Gao Ke Xi

House No. 1   Winter

Time	SITES					
	1	2	3	4	5	6
	RH%	RH%	RH%	RH%	RH%	RH%
8:00 pm	76.9	79.8	74.4	77.5	80.1	85.8
9:00	79.5	79.8	- 74.3	70.6	81.3	88.8
10:00	80.1	83.6	74.4	72.4	83.6	85.7
11:00	80.1	81.8	72.6	71.9	80.8	85.7
Midnight	80.8	83.1	68.7	72.0	83.6	85.7
1:00 am	80.2	80.7	70.5	76.4	89.8	88.7
2:00	79.6	77.1	71.6	76.4	94.4	88.7
3:00	79.6	79.0	75.3	76.9	89.6	93.4
4:00	79.1	80.2	76.6	77.9	88.7	93.4
5:00	78.6	80.2	76.6	77.9	88.7	88.7
6:00	76.9	77.9	74.9	77.9	85.7	88.7
7:00	75.7	77.9	71.4	73.5	90.2	87.2
8:00	73.4	77.4	74.3	72.4	95.0	85.8
9:00	66.7	68.3	66.7	62.6	91.2	88.7
10:00	72.6	78.3	68.9	68.9	92.0	91.9
11:00	71.5	68.3	67.4	69.0	86.2	86.1
Noon	72.0	71.2	69.6	73.3	89.4	88.3
1:00 pm	70.2	74.3	70.5	72.9	89.6	87.8
2:00	73.9	70.6	68.9	67.8	89.6	91.6
3:00	72.7	77.1	69.1	70.8	81.7	82.2
4:00	72.6	80.2	68.3	68.9	86.3	86.2
5:00	73.3	79.4	67.7	74.0	89.1	86.1
6:00	80.4	80.7	71.0	65.7	88.8	88.1
7:00	84.8	84.7	74.6	57.1	88.8	91.9
8:00	82.5	85.5	77.9	67.0	95.0	88.7

Diurnal Temperature Measurements

COUNTY: Yan'an Region
PROVINCE: Shaanxi
SETTLEMENT NAME: Wang Jia Terrace
FAMILY/HOUSE NAME: Gou Shengzji

House No. 2   Winter
Jan8-9, 1985
Degrees in Centigrade

Time	SITES											
	1		2		3		4		5		6	
	Dry	Wet	Dry	Wet	Dry	Wet	Dry	Wet	Dry	Wet	Dry	Wet
8:00 pm	11.7	8.0	11.2	7.4	12.0	7.9	-6.2	-7.2				
9:00	10.7	6.2	11.7	7.9	12.2	7.9	-7.8	-8.8				
10:00	10.2	5.8	11.7	8.7	11.2	6.8	-8.7	-9.0				
11:00	10.0	5.7	11.2	7.8	10.7	6.4	-10.7	-11.3				
Midnight	9.8	5.7	10.6	7.5	10.4	6.0	-11.0	-11.7				
1:00 am	9.5	5.7	10.1	7.2	10.0	5.6	-11.2	-12.0				
2:00	9.3	5.4	9.6	6.7	9.5	5.3	-11.7	-12.5				
3:00	9.1	4.7	9.0	6.2	9.2	4.9	-12.1	-13.0				
4:00	8.9	4.5	8.7	5.8	9.0	4.5	-13.1	-14.1				
5:00	8.8	4.4	8.4	5.5	8.8	4.0	-14.1	-15.1				
6:00	7.9	4.2	8.3	5.2	8.4	3.7	-14.3	-15.5				
7:00	8.3	3.9	8.5	5.9	8.6	4.5	-14.6	-15.6				
8:00	9.3	6.7	10.1	7.8	9.1	5.7	-15.0	-16.1				
9:00	8.9	4.5	10.7	8.4	10.6	7.6	-14.0	-15.0				
10:00	8.9	5.3	10.1	6.6	10.9	7.2	-12.0	-12.7				
11:00	9.4	5.7	9.3	6.6	11.2	7.4	-9.0	-10.0				
Noon	9.4	4.9	9.1	5.6	12.4	7.2	-4.3	-6.7				
1:00 pm	12.4	8.3	10.0	6.3	12.8	9.2	-2.3	-5.2				
2:00	13.4	8.5	10.1	6.2	13.0	8.0	-2.0	-4.4				
3:00	13.7	8.2	11.0	7.0	12.8	8.6	-1.7	-4.4				
4:00	13.4	7.4	10.7	6.0	12.6	7.8	-1.7	-4.1				
5:00	12.2	6.6	10.4	6.1	12.6	7.4	-2.3	-5.3				
6:00	11.0	5.6	10.6	7.7	13.1	8.5	-4.7	-6.7				
7:00	10.8	7.4	11.4	8.3	13.0	8.4	-6.3	-8.2				
8:00	11.2	6.7	11.7	8.0	12.9	8.3	-8.0	-8.7				

Relative Humidity Calculations

COUNTY: Yan'an Region
PROVINCE: Shaanxi
SETTLEMENT NAME: Wang Tia Terrace
FAMILY/HOUSE NAME: Gou Shengzhi

House No. 2   Summer

Time	SITES 1 RH%	2 RH%	3 RH%	4 RH%	5 RH%	6 RH%
8:00 pm	77.9	84.7	82.1	96.9		
9:00	79.0	83.0	84.6	98.9		
10:00	77.8	83.4	89.8	98.4		
11:00	79.4	80.9	91.1	100.0		
Midnight	80.7	82.3	83.2	98.9		
1:00 am	81.3	81.3	83.6	99.4		
2:00	80.4	82.1	83.9	97.9		
3:00	82.4	83.0	83.1	98.4		
4:00	81.9	84.2	83.5	98.9		
5:00	84.1	86.4	84.7	98.4		
6:00	83.2	82.1	84.7	98.9		
7:00	82.7	78.7	83.4	96.4		
8:00	84.8	80.3	84.3	92.6		
9:00	80.0	82.9	82.8	97.0		
10:00	82.5	82.9	84.4	93.7		
11:00	81.7	82.5	86.0	92.4		
Noon	83.0	84.3	86.1	88.9		
1:00 pm	81.3	82.1	83.2	80.1		
2:00	82.8	80.0	83.2	89.4		
3:00	78.8	83.0	85.7	88.5		
4:00	82.3	79.1	81.7	88.9		
5:00	83.0	79.9	81.3	88.3		
6:00	82.2	79.9	82.9	91.4		
7:00	82.0	78.7	80.4	93.7		
8:00	79.6	78.7	79.8	93.6		

Diurnal Temperature Measurements

COUNTY: Yan'an Region
PROVINCE: Shaanxi
SETTLEMENT NAME: Gao Me Wan Village
FAMILY/HOUSE NAME: Zhang Yen Fu

House No. 3   Winter
Jan. 10-11, 1985
Degrees in Centigrade

Time	1 Dry	1 Wet	2 Dry	2 Wet	3 Dry	3 Wet	4 Dry	4 Wet	5 Dry	5 Wet	6 Dry	6 Wet
8:00 pm	0.3	-3.0	-1.1	-4.5	10.1	8.5	6.8	5.6	-5.4	-5.9		
9:00	-0.5	-3.8	-2.2	-5.2	9.7	7.9	6.3	4.4	-6.1	-7.2		
10:00	-1.1	-4.8	-2.8	-5.5	9.5	7.8	5.8	4.4	-7.2	-8.0		
11:00	-1.5	-5.1	-3.5	-6.1	9.0	6.7	5.5	3.4	-8.0	-8.8		
Midnight	-1.9	-5.9	-4.1	-6.5	8.4	6.2	5.2	3.7	-8.3	-9.7		
1:00 am	-2.0	-5.6	-4.2	-6.3	8.2	5.8	5.1	3.5	-8.2	-9.5		
2:00	-2.2	-5.4	-4.2	-6.2	7.9	5.4	4.9	3.3	-8.0	-9.1		
3:00	-2.3	-5.5	-4.5	-6.6	7.6	5.1	4.8	3.1	-8.6	-9.7		
4:00	-2.6	-5.5	-4.7	-7.2	7.3	4.8	4.5	2.8	-9.0	-10.2		
5:00	-2.7	-5.5	-5.0	-7.2	7.1	4.6	4.4	2.6	-9.2	-10.2		
6:00	-2.7	-5.7	-5.2	-7.1	6.8	4.4	4.2	2.4	-9.6	-10.3		
7:00	-2.8	-5.9	-5.3	-7.2	6.6	4.1	4.1	2.3	-9.7	-10.5		
8:00	-2.7	-5.9	-5.0	-7.2	6.7	4.2	4.2	2.8	-9.3	-10.2		
9:00	-2.7	-5.7	-5.0	-6.9	7.0	4.5	4.9	3.0	-8.6	-9.6		
10:00	-2.6	-5.0	-5.0	-6.6	8.7	6.1	7.2	5.2	-7.8	-8.3		
11:00	-2.2	-5.0	-4.4	-6.1	8.3	5.6	6.4	4.9	-5.2	-6.2		
Noon	-1.8	-4.9	-3.8	-5.8	7.8	5.2	5.7	4.5	-2.2	-3.7		
1:00 pm	-1.1	-3.8	-3.3	-5.0	7.9	5.3	6.2	4.2	-1.7	-2.8		
2:00	-0.2	-3.6	-2.8	-4.4	8.4	5.1	7.0	5.1	-1.1	-2.0		
3:00	0.2	-3.0	-1.7	-4.0	9.5	5.2	6.4	4.5	0.1	-3.2		
4:00	0.1	-2.7	-1.7	-3.8	9.1	6.3	6.2	4.4	-0.7	-3.7		
5:00	0.1	-3.3	-1.8	-4.0	8.8	6.7	5.7	3.9	-1.7	-4.7		
6:00	-0.3	-3.5	-2.1	-4.3	8.5	6.3	5.9	5.1	-3.6	-4.7		
7:00	-0.5	-3.6	-2.2	-4.3	8.7	7.2	6.6	5.8	-4.1	-4.7		
8:00	-0.8	-3.8	-2.2	-4.4	8.9	7.3	6.5	5.7	-5.2	-5.9		

Diurnal Temperature Measurements

COUNTY: Yan'an Region
PROVINCE: Shaanxi
SETTLEMENT NAME: Gao Me Wan Village
FAMILY/HOUSE NAME: Zhang Yen Fu

House No. 3   Summer
July 19-20, 1984
Degrees in Centigrade

Time	1 Dry	1 Wet	2 Dry	2 Wet	3 Dry	3 Wet	4 Dry	4 Wet	5 Dry	5 Wet	6 Dry	6 Wet
8:00 pm	20.0	17.4	20.1	17.8	23.1	20.8	23.4	20.3	19.8	18.9		
9:00	19.5	17.2	19.8	17.8	22.9	20.7	23.4	21.2	19.4	18.4		
10:00	19.3	17.0	19.7	17.8	23.0	20.3	23.4	20.9	19.2	18.3		
11:00	19.1	16.9	19.5	17.8	22.9	20.2	23.3	20.3	18.9	18.3		
Midnight	19.0	16.7	19.5	17.7	22.7	20.3	23.1	20.9	18.6	17.8		
1:00 am	19.0	16.7	19.4	17.6	22.7	20.5	23.2	20.9	18.4	17.7		
2:00	18.9	16.8	19.5	17.6	22.7	20.2	23.1	20.9	17.3	16.8		
3:00	18.9	16.7	19.5	17.6	22.6	19.8	22.8	19.8	17.3	16.4		
4:00	18.9	16.7	19.5	17.4	22.3	19.6	22.6	19.6	16.6	16.2		
5:00	18.9	16.7	19.4	17.3	22.2	19.4	22.3	19.5	16.1	15.6		
6:00	18.8	16.6	19.4	17.3	21.7	18.5	22.3	19.0	15.6	15.6		
7:00	18.8	16.4	19.2	17.2	21.8	19.2	22.8	20.1	17.8	16.4		
8:00	18.9	16.4	19.2	17.3	21.8	18.9	22.9	19.8	19.5	17.3		
9:00	18.9	16.8	19.2	17.0	21.9	18.8	23.0	19.7	19.5	17.7		
10:00	18.9	16.7	19.2	17.3	22.3	19.0	23.1	19.2	20.6	18.4		
11:00	19.1	17.0	19.4	17.3	23.1	20.2	24.0	20.8	24.4	19.0		
Noon	19.5	17.3	19.6	17.8	25.0	21.1	24.7	21.4	24.2	20.2		
1:00 pm	19.6	17.4	19.8	17.9	24.8	21.1	24.8	21.7	23.9	20.5		
2:00	19.8	17.8	20.3	18.5	24.8	21.7	25.1	21.7	27.8	21.7		
3:00	20.6	18.7	20.9	19.5	24.6	21.4	24.9	21.7	27.0	22.3		
4:00	20.6	18.7	20.8	19.5	25.0	21.7	25.1	22.0	27.6	22.3		
5:00	20.7	18.7	20.8	19.0	24.7	21.8	24.8	22.2	27.3	22.8		
6:00	20.7	18.5	20.7	19.0	24.2	21.6	24.6	21.7	26.4	22.8		
7:00	20.3	18.3	20.6	18.7	24.1	21.3	24.4	21.4	23.4	20.9		
8:00	20.1	18.1	20.9	18.7	23.7	21.1	23.9	21.2	22.3	20.1		

Relative Humidity Calculations

COUNTY: Yan'an Region
PROVINCE: Shaanxi
SETTLEMENT NAME: Gao Me Wan Village
FAMILY/HOUSE NAME: Zhang Yen Fu

House No. 3   Summer

Time	1 RH%	2 RH%	3 RH%	4 RH%	5 RH%	6 RH%
8:00 pm	80.3	83.1	83.7	78.2	93.4	
9:00	82.1	85.0	84.4	84.1	92.1	
10:00	82.4	85.4	81.0	82.3	92.5	
11:00	83.2	86.6	80.6	79.3	95.0	
Midnight	81.9	86.2	82.8	84.0	93.7	
1:00 am	82.3	85.3	83.9	83.7	94.5	
2:00	83.1	84.9	81.7	84.0	95.8	
3:00	82.7	84.9	80.1	78.3	93.1	
4:00	82.7	83.7	80.7	78.6	104.9	
5:00	83.1	83.3	79.6	79.6	104.9	
6:00	82.2	83.7	76.8	77.0	100.0	
7:00	81.4	84.8	80.9	80.6	88.7	
8:00	81.0	84.9	79.4	78.0	82.9	
9:00	83.1	82.8	77.6	76.6	86.2	
10:00	82.7	84.9	76.6	72.7	83.3	
11:00	83.6	83.3	79.6	77.4	64.6	
Noon	82.9	85.8	73.9	78.0	72.6	
1:00 pm	83.3	85.8	75.2	79.1	76.3	
2:00	84.6	86.4	78.7	77.1	62.7	
3:00	85.7	89.4	78.3	77.7	69.9	
4:00	85.3	90.2	77.4	79.2	67.4	
5:00	84.5	86.9	80.1	82.0	71.7	
6:00	83.7	87.7	81.8	80.1	76.3	
7:00	84.8	85.7	80.7	79.3	82.3	
8:00	85.1	83.4	81.3	80.6	83.8	

Relative Humidity Calculations

COUNTY: Yan'an Region
PROVINCE: Shaanxi
SETTLEMENT NAME: Gao Me Wan Village
FAMILY/HOUSE NAME: Zhang Yen Fu

House No. 3   Winter

Time	SITES					
	1	2	3	4	5	6
	RH%	RH%	RH%	RH%	RH%	RH%
8:00 pm	49.4	45.7	82.3	84.9	88.2	
9:00	47.3	48.1	80.8	75.8	75.6	
10:00	39.9	51.3	80.2	82.2	83.0	
11:00	39.6	52.4	74.4	73.7	78.2	
Midnight	32.1	53.9	73.9	81.1	65.4	
1:00 am	38.8	57.9	72.4	80.3	68.4	
2:00	44.6	62.0	70.9	78.1	71.5	
3:00	45.9	57.3	70.6	77.2	70.4	
4:00	48.9	50.4	69.7	77.0	68.3	
5:00	49.7	55.0	70.1	76.1	73.6	
6:00	46.7	61.1	70.4	75.3	78.9	
7:00	43.7	60.9	68.9	75.8	77.3	
8:00	43.0	56.0	69.0	80.4	77.7	
9:00	46.7	61.5	69.4	74.5	74.6	
10:00	58.3	67.1	71.0	75.9	86.4	
11:00	52.7	65.9	69.4	80.6	79.1	
Noon	48.2	62.8	69.0	84.9	72.5	
1:00 pm	55.2	67.9	69.7	75.1	79.7	
2:00	47.9	70.6	62.4	76.4	84.9	
3:00	50.7	62.0	53.7	76.6	50.2	
4:00	56.1	64.7	69.1	78.4	51.8	
5:00	48.6	63.6	76.0	77.3	40.3	
6:00	49.5	61.4	74.5	89.3	77.7	
7:00	50.6	63.0	82.1	90.2	87.8	
8:00	50.8	62.0	81.6	80.5	83.7	

Diurnal Temperature Measurements

COUNTY: Gao Lan
PROVINCE: Gansu
SETTLEMENT NAME: Ya Chuan Village
FAMILY/HOUSE NAME: Cao Yiren

House No. 4   Winter
Dec.14-15,1984
Degrees in Centigrade

Time	SITES											
	1		2		3		4		5		6	
	Dry	Wet	Dry	Wet	Dry	Wet	Dry	Wet	Dry	Wet	Dry	Wet
8:00 pm	11.2	8.4	10.9	8.2	9.5	7.3	2.3	-1.1	-6.7	-7.6		
9:00	11.1	8.4	11.1	8.7	9.9	7.4	1.9	-1.1	-7.2	-8.0		
10:00	10.8	8.4	11.3	8.6	10.1	8.0	1.8	-1.6	-7.3	-8.2		
11:00	10.7	8.1	11.3	8.8	10.1	7.5	1.7	-1.6	-7.6	-8.6		
Midnight	10.7	8.7	11.3	8.6	10.1	8.1	1.7	-1.9	-8.0	-8.7		
1:00 am	10.7	8.1	11.2	8.5	10.1	7.4	1.6	-2.0	-8.6	-8.7		
2:00	10.7	8.2	11.2	8.1	9.9	7.4	1.6	-2.0	-8.3	-8.7		
3:00	10.7	8.1	11.2	8.1	10.1	7.3	1.3	-2.1	-8.6	-9.0		
4:00	10.6	7.9	11.2	7.9	10.1	7.3	0.7	-2.7	-8.6	-9.2		
5:00	10.6	7.9	11.2	7.7	9.9	7.3	1.2	-2.3	-8.6	-9.2		
6:00	10.6	7.9	11.2	7.7	9.9	7.3	1.2	-2.2	-8.6	-9.2		
7:00	10.6	7.9	11.3	7.9	10.1	7.3	1.2	-2.2	-8.6	-9.2		
8:00	10.7	8.1	11.2	7.9	10.1	7.4	1.2	-2.2	-8.7	-9.2		
9:00	11.1	7.9	11.4	8.1	9.5	7.2	1.2	-2.1	-8.0	-8.7		
10:00	11.3	8.9	11.3	9.2	10.1	7.7	1.2	-2.2	-8.2	-8.7		
11:00	11.3	8.9	11.3	9.0	10.1	7.7	1.3	-2.1	-7.0	-8.0		
Noon	10.7	7.3	10.2	8.4	8.4	5.6	1.6	-2.0	-6.1	-7.3		
1:00 pm	11.1	8.9	11.2	8.3	9.0	6.7	2.0	-1.1	-4.0	-5.0		
2:00	11.3	8.9	11.2	8.5	9.5	7.2	2.2	-0.9	-4.1	-5.2		
3:00	11.7	8.6	11.7	8.4	9.6	7.3	2.2	-0.7	-2.3	-5.3		
4:00	11.3	8.5	11.4	8.8	9.9	7.7	2.3	-0.9	-4.6	-5.5		
5:00	11.3	7.3	11.3	7.6	9.0	6.2	2.2	-0.9	-5.5	-6.2		
6:00	11.2	8.5	11.6	8.5	9.3	6.7	1.9	-1.6	-6.2	-7.2		
7:00	11.0	8.1	11.2	7.9	10.1	6.6	1.7	-2.5	-7.7	-9.0		
8:00	11.8	9.4	12.6	9.9	10.1	8.1	1.7	-1.9	-8.8	-9.8		

Diurnal Temperature Measurements

COUNTY: Gao Lan
PROVINCE: Gansu
SETTLEMENT NAME: Ya Chuan Village
FAMILY/HOUSE NAME: Cao Yiren

House No. 4   Summer
July26-27,1984
Degrees in Centigrade

Time	SITES											
	1		2		3		4		5		6	
	Dry	Wet	Dry	Wet	Dry	Wet	Dry	Wet	Dry	Wet	Dry	Wet
8:00 pm	22.3	15.2	22.4	16.3	18.1	16.3	18.8	14.0	21.8	14.6		
9:00	21.6	15.1	22.1	15.6	18.3	16.3	18.5	14.4	19.4	14.2		
10:00	21.7	15.9	21.9	16.2	18.5	16.6	18.4	14.6	22.8	14.6		
11:00	21.7	15.9	21.8	15.9	18.5	16.6	18.3	14.5	18.9	14.0		
Midnight	20.6	15.5	21.2	15.7	18.1	15.9	18.1	14.1	18.3	13.7		
1:00 am	20.1	15.3	20.6	15.1	17.6	15.7	17.9	14.1	14.6	12.8		
2:00	19.5	15.1	19.8	15.0	17.3	15.1	17.8	13.9	13.9	12.3		
3:00	19.0	14.5	19.2	13.9	16.8	14.9	17.6	13.5	12.8	11.4		
4:00	18.5	13.9	18.4	12.9	16.6	14.6	17.3	13.4	12.0	10.9		
5:00	18.3	13.8	17.9	12.8	16.3	14.5	17.0	13.1	11.8	10.6		
6:00	17.8	13.1	17.4	12.6	16.1	14.1	16.7	12.9	10.6	10.2		
7:00	17.5	13.6	17.3	12.8	15.9	14.0	16.4	12.9	16.2	12.6		
8:00	18.3	14.1	18.4	14.5	16.3	14.6	16.8	13.6	22.0	14.5		
9:00	19.2	15.3	19.8	15.5	17.2	15.6	17.4	14.6	27.2	16.3		
10:00	20.1	15.6	20.6	16.7	17.8	16.3	18.1	14.5	29.2	18.4		
11:00	20.7	15.9	21.8	17.3	18.7	17.2	18.5	14.8	30.1	18.7		
Noon	21.7	16.2	22.5	17.8	19.2	17.3	19.0	15.2	30.2	19.2		
1:00 pm	22.2	16.2	23.4	18.3	19.7	17.4	19.7	15.5	30.1	19.5		
2:00	22.1	16.4	23.4	17.3	20.0	16.8	20.1	15.7	30.7	19.8		
3:00	21.7	16.4	23.1	16.8	19.9	16.8	20.5	16.2	33.3	18.7		
4:00	22.0	16.4	23.1	17.0	19.8	17.5	20.7	16.3	29.5	18.1		
5:00	22.8	16.7	24.4	17.3	19.8	17.9	20.7	15.7	29.6	17.6		
6:00	25.6	17.5	24.6	17.3	20.7	17.7	20.7	16.4	28.4	18.4		
7:00	23.9	17.3	24.2	18.3	20.6	18.1	20.6	15.9	27.7	18.7		
8:00	23.4	17.9	24.3	18.4	20.6	18.3	20.5	16.2	26.7	18.1		

Relative Humidity Calculations

COUNTY: Gao Lan
PROVINCE: Gansu
SETTLEMENT NAME: Ya Chuan Village
FAMILY/HOUSE NAME: Cao Yiren

House No. 4   Summer

Time	SITES					
	1	2	3	4	5	6
	RH%	RH%	RH%	RH%	RH%	RH%
8:00 pm	49.1	55.9	84.5	61.2	47.9	
9:00	51.9	53.1	82.7	66.0	59.0	
10:00	56.7	57.3	83.7	68.5	43.0	
11:00	57.0	56.1	83.7	68.4	60.2	
Midnight	60.6	58.4	80.8	66.5	62.0	
1:00 am	62.8	57.5	83.3	67.3	83.0	
2:00	64.4	62.2	80.4	67.2	83.7	
3:00	63.2	58.0	83.0	65.6	85.9	
4:00	62.2	56.0	81.9	66.7	88.4	
5:00	63.3	57.9	83.7	66.5	87.2	
6:00	61.1	59.4	82.2	67.1	95.1	
7:00	66.9	62.3	82.1	69.1	68.0	
8:00	65.0	67.6	84.2	72.2	46.4	
9:00	68.3	65.8	86.0	75.4	34.3	
10:00	64.8	69.3	86.3	70.0	37.0	
11:00	63.0	65.5	87.0	69.0	35.4	
Noon	58.6	65.2	83.5	68.2	37.5	
1:00 pm	56.1	62.9	80.6	66.5	39.2	
2:00	58.1	56.2	74.3	64.9	38.1	
3:00	60.4	55.2	74.7	66.0	25.9	
4:00	58.8	56.3	81.1	65.7	34.8	
5:00	56.0	51.4	84.2	61.8	32.1	
6:00	46.9	50.3	76.4	66.9	40.0	
7:00	53.9	58.9	80.1	63.3	44.1	
8:00	60.1	58.7	81.0	66.4	45.3	

Relative Humidity Calculations

COUNTY: Gao Lan
PROVINCE: Gansu
SETTLEMENT NAME: Ya Chuan Village
FAMILY/HOUSE NAME: Cao Yiren

House No. 4   Winter

Time	SITES 1 RH%	2 RH%	3 RH%	4 RH%	5 RH%	6 RH%
8:00 pm	71.7	71.5	74.8	53.9	81.0	
9:00	72.2	74.4	72.2	58.3	81.7	
10:00	74.7	72.3	77.6	52.6	80.3	
11:00	72.4	74.5	72.3	53.2	76.1	
Midnight	78.0	71.8	78.7	49.4	80.9	
1:00 am	72.4	72.2	71.2	49.1	95.7	
2:00	74.0	68.4	72.2	49.7	90.1	
3:00	72.3	67.9	70.6	51.2	90.0	
4:00	71.7	66.3	70.1	49.7	84.4	
5:00	71.7	64.6	71.1	50.2	84.4	
6:00	71.7	64.6	71.1	50.3	84.4	
7:00	71.7	65.8	70.1	50.9	84.4	
8:00	72.3	66.3	71.8	51.7	88.5	
9:00	67.2	66.0	74.8	52.5	80.9	
10:00	74.6	77.9	74.1	51.1	86.1	
11:00	75.6	75.7	74.7	51.4	76.9	
Noon	64.6	80.6	68.3	50.5	73.2	
1:00 pm	77.7	70.6	73.8	56.9	80.4	
2:00	75.1	72.2	74.8	57.4	79.2	
3:00	67.8	66.3	74.3	61.1	47.8	
4:00	71.2	73.0	75.7	56.1	80.9	
5:00	59.4	62.6	68.4	57.5	84.6	
6:00	72.2	68.3	71.6	52.1	77.9	
7:00	69.9	66.2	62.3	42.0	70.6	
8:00	75.5	72.9	78.7	50.0	74.2	

Diurnal Temperature Measurements

COUNTY: Yang Qu
PROVINCE: Shanxi
SETTLEMENT NAME: Qing Long Village
FAMILY/HOUSE NAME: Zhao Qingyu

House No. 5   Winter
Dec.22-23,1984
Degrees in Centigrade

Time	1 Dry	1 Wet	2 Dry	2 Wet	3 Dry	3 Wet	4 Dry	4 Wet	5 Dry	5 Wet	6 Dry	6 Wet
8:00 pm	-6.1	-9.4	9.9	5.7	9.5	6.3	-15.0	-16.0	-15.2	-16.1		
9:00	-7.2	-9.6	9.7	6.2	9.5	6.1	-15.7	-16.5	-16.0	-16.5		
10:00	-7.6	-10.0	9.8	5.9	9.3	6.1	-16.5	-16.6	-16.6	-16.5		
11:00	-7.7	-10.0	9.5	5.7	8.9	5.4	-16.4	-16.5	-16.6	-16.5		
Midnight	-7.7	-10.1	9.4	5.6	8.7	5.2	-16.5	-16.6	-16.6	-16.5		
1:00 am	-8.3	-10.5	9.2	5.7	8.6	5.3	-16.6	-16.6	-16.6	-16.6		
2:00	-8.8	-11.0	9.1	5.6	8.6	5.6	-16.6	-16.6	-16.6	-16.6		
3:00	-9.3	-11.6	8.8	5.4	8.5	5.4	-16.6	-16.6	-16.6	-16.6		
4:00	-10.0	-12.1	8.6	5.1	8.4	5.2	-16.6	-16.6	-16.6	-16.6		
5:00	-10.3	-12.7	8.5	5.1	8.1	5.0	-16.6	-16.6	-16.6	-16.6		
6:00	-10.8	-13.3	8.4	5.0	7.8	4.8	-16.6	-16.6	-16.6	-16.6		
7:00	-11.3	-13.6	8.4	4.8	7.7	4.4	-16.6	-16.6	-16.6	-16.6		
8:00	-11.9	-14.2	8.4	4.5	7.6	4.1	-16.5	-16.6	-16.6	-16.6		
9:00	-11.6	-13.8	8.4	4.3	7.4	4.2	-16.5	-16.5	-16.6	-16.6		
10:00	-9.4	-10.7	9.2	5.7	9.0	5.7	-16.2	-16.5	-16.5	-16.5		
11:00	-7.8	-9.3	9.5	6.2	10.1	6.5	-13.6	-14.6	-14.1	-15.0		
Noon	-6.2	-7.9	10.9	7.3	11.8	7.2	-11.5	-12.8	-12.6	-13.7		
1:00 pm	-5.0	-6.6	11.6	8.4	13.0	8.3	-9.6	-11.0	-10.3	-11.3		
2:00	-4.3	-5.9	11.7	7.3	12.7	7.9	-8.3	-9.7	-9.7	-10.6		
3:00	-3.9	-5.8	11.9	7.7	11.7	6.8	-8.3	-9.6	-9.2	-10.2		
4:00	-4.2	-6.6	11.6	7.3	11.2	6.8	-8.2	-9.0	-8.8	-9.7		
5:00	-4.8	-7.2	10.8	6.6	9.8	6.2	-10.6	-11.3	-11.2	-12.0		
6:00	-5.5	-7.8	10.2	5.6	9.1	5.9	-11.8	-13.6	-13.5	-14.7		
7:00	-6.6	-8.9	9.8	5.6	8.9	5.6	-13.7	-15.0	-14.5	-15.7		
8:00	-6.6	-9.0	9.8	5.7	8.7	5.5	-14.8	-15.7	-15.2	-16.5		

Diurnal Temperature Measurements

COUNTY: Yang Qu
PROVINCE: Shanxi
SETTLEMENT NAME: Qing Long Village
FAMILY/HOUSE NAME: Zhao Qingyu

House No. 5   Summer
Aug.31-Sep.1,1984
Degrees in Centigrade

Time	1 Dry	1 Wet	2 Dry	2 Wet	3 Dry	3 Wet	4 Dry	4 Wet	5 Dry	5 Wet	6 Dry	6 Wet
8:00 pm	18.2	14.5	21.7	17.3	21.1	16.8	21.0	14.6	21.7	15.1		
9:00	17.8	15.6	22.0	18.6	21.2	18.4	20.6	16.2	20.9	16.3		
10:00	17.8	16.1	21.8	18.9	21.4	19.3	19.0	16.2	20.1	16.2		
11:00	17.8	15.9	21.7	18.9	21.4	19.0	18.8	16.2	19.8	15.9		
Midnight	17.8	15.8	21.7	18.9	21.3	19.2	18.7	15.8	19.0	15.6		
1:00 am	17.8	15.6	21.7	19.1	21.2	19.4	18.4	15.8	18.1	15.6		
2:00	17.8	16.2	21.6	19.0	21.2	19.5	17.8	16.4	17.8	16.1		
3:00	17.6	16.2	21.6	19.1	21.1	19.0	16.3	15.2	17.4	15.1		
4:00	17.4	15.6	21.3	18.7	21.1	18.5	15.7	15.5	15.6	14.1		
5:00	17.3	15.6	21.3	18.9	20.7	18.4	15.6	15.1	15.6	14.5		
6:00	17.2	15.6	21.2	18.4	20.7	17.9	15.3	14.8	15.7	14.6		
7:00	17.3	15.7	21.2	18.4	20.6	17.8	16.2	15.6	16.7	15.1		
8:00	17.4	15.9	21.1	18.4	20.6	17.4	18.4	16.7	17.8	15.1		
9:00	17.5	16.1	20.9	18.2	20.6	18.5	18.4	16.2	18.5	15.7		
10:00	17.6	16.2	21.2	18.9	20.6	18.1	25.1	18.5	24.5	18.1		
11:00	17.8	16.2	21.6	17.9	21.3	18.9	27.6	20.2	27.8	19.0		
Noon	18.3	16.7	22.4	18.8	21.9	19.1	29.6	20.9	25.6	18.9		
1:00 pm	18.4	16.7	22.2	18.4	21.8	19.6	27.9	20.9	26.2	18.9		
2:00	18.4	16.4	22.3	18.4	21.8	19.0	26.2	17.9	26.8	19.2		
3:00	18.4	16.4	22.6	18.3	22.2	18.4	26.3	17.3	26.4	18.9		
4:00	18.4	16.2	22.3	17.8	21.9	18.4	23.9	16.3	23.4	17.0		
5:00	18.4	15.6	21.7	16.2	21.4	17.9	20.1	15.9	19.8	15.3		
6:00	18.1	15.7	21.2	17.6	20.9	18.0	18.2	15.8	18.4	15.3		
7:00	17.9	16.4	21.3	18.1	20.9	17.9	16.1	15.5	17.8	14.2		
8:00	17.8	15.6	21.3	18.4	20.6	17.4	14.2	13.9	15.1	13.7		

Relative Humidity Calculations

COUNTY: Yang Qu
PROVINCE: Shanxi
SETTLEMENT NAME: Qing Long Village
FAMILY/HOUSE NAME: Zhao Qingyu

House No. 5   Summer

Time	1 RH%	2 RH%	3 RH%	4 RH%	5 RH%	6 RH%
8:00 pm	68.8	66.2	67.2	51.9	51.3	
9:00	80.6	73.4	78.3	65.3	64.3	
10:00	84.8	77.8	82.6	75.9	68.9	
11:00	83.0	77.7	81.3	78.4	68.7	
Midnight	82.0	78.2	83.4	75.7	71.9	
1:00 am	81.1	79.4	85.6	77.3	78.5	
2:00	85.3	79.8	86.0	87.7	85.3	
3:00	87.6	80.2	82.9	89.7	79.0	
4:00	83.7	78.8	79.5	98.4	85.9	
5:00	85.1	81.3	81.5	95.2	89.5	
6:00	85.5	77.9	77.6	94.6	90.0	
7:00	86.1	77.9	77.6	94.2	84.9	
8:00	87.1	78.7	74.2	85.5	76.1	
9:00	87.6	78.2	82.7	81.4	76.4	
10:00	87.1	82.1	79.7	55.0	55.8	
11:00	85.3	71.6	81.3	52.8	45.3	
Noon	86.0	72.5	77.9	46.8	55.2	
1:00 pm	85.5	71.5	82.0	54.9	52.4	
2:00	82.7	70.4	78.2	46.6	51.1	
3:00	82.3	67.9	71.5	42.8	51.1	
4:00	81.4	66.6	73.4	48.2	54.8	
5:00	76.4	58.9	72.3	66.9	64.6	
6:00	78.9	71.7	76.5	79.0	74.2	
7:00	87.2	75.0	76.5	94.2	69.4	
8:00	80.6	77.1	74.2	97.2	86.8	

Relative Humidity Calculations

COUNTY: Yang Qu
PROVINCE: Shanxi
SETTLEMENT NAME: Qing Long Village
FAMILY/HOUSE NAME: Zhao Qingyu

House No. 5 — Winter

Time	SITES					
	1 RH%	2 RH%	3 RH%	4 RH%	5 RH%	6 RH%
8:00 pm	29.3	55.9	64.9	61.0	62.5	
9:00	44.0	61.8	63.2	70.1	78.9	
10:00	44.0	58.5	65.2	95.1	100.0	
11:00	44.9	58.7	60.8	95.1	100.0	
Midnight	44.7	58.6	61.0	92.6	100.0	
1:00 am	45.4	61.1	63.2	100.0	100.0	
2:00	46.3	62.1	66.1	100.0	100.0	
3:00	40.3	62.3	64.9	100.0	100.0	
4:00	40.9	60.9	63.0	100.0	100.0	
5:00	33.3	61.3	64.4	97.5	100.0	
6:00	28.2	61.2	64.7	97.5	100.0	
7:00	33.6	58.9	62.6	100.0	100.0	
8:00	29.4	57.1	60.0	95.1	97.5	
9:00	32.3	54.7	61.7	100.0	97.5	
10:00	65.9	61.7	63.7	85.6	100.0	
11:00	62.6	64.3	61.7	64.7	67.4	
Noon	62.4	63.2	54.9	57.8	61.7	
1:00 pm	67.1	67.2	55.6	62.6	71.6	
2:00	68.3	55.9	54.2	65.4	75.8	
3:00	63.7	58.1	51.3	69.5	75.0	
4:00	52.8	57.2	55.7	80.6	75.7	
5:00	52.1	56.6	61.8	77.5	76.6	
6:00	51.5	52.5	64.9	46.5	57.3	
7:00	48.4	55.7	63.1	54.7	52.1	
8:00	46.0	56.3	64.0	65.6	51.5	

Diurnal Temperature Measurements

COUNTY: Linfen City
PROVINCE: Shanxi
SETTLEMENT NAME: Da Yang Village
FAMILY/HOUSE NAME: Cui Mingxing

House No. 6 — Winter
Dec.19-20,1984
Degrees in Centigrade

Time	1 Dry	1 Wet	2 Dry	2 Wet	3 Dry	3 Wet	4 Dry	4 Wet	5 Dry	5 Wet	6 Dry	6 Wet
8:00 pm	13.3	9.9	12.8	9.1	12.5	9.6	11.3	7.9	-5.7	-6.7		
9:00	13.4	9.6	13.0	9.2	12.9	9.4	11.4	7.7	-6.7	-7.7		
10:00	13.4	10.0	13.1	9.5	13.0	9.7	11.4	8.1	-7.2	-8.0		
11:00	13.4	10.0	13.1	9.5	12.9	9.5	11.3	7.9	-7.6	-8.7		
Midnight	13.4	9.7	13.0	9.1	12.6	8.9	11.3	7.9	-8.2	-9.0		
1:00 am	13.4	9.7	12.9	8.9	12.4	8.9	11.2	7.9	-8.6	-9.5		
2:00	13.4	9.7	12.9	8.7	12.3	8.9	11.1	7.9	-9.0	-10.0		
3:00	13.4	9.5	12.9	8.8	12.3	8.9	11.0	7.7	-9.6	-10.7		
4:00	13.4	9.2	12.9	8.8	12.3	8.9	10.9	7.4	-10.1	-11.2		
5:00	13.4	9.2	12.9	8.9	12.3	8.9	10.9	7.3	-10.1	-11.2		
6:00	13.4	9.3	12.8	8.9	12.1	8.9	10.8	7.1	-10.0	-11.0		
7:00	13.4	9.2	12.8	8.9	12.1	8.9	10.7	7.2	-9.5	-10.6		
8:00	13.4	9.5	12.8	8.9	12.3	8.4	10.9	7.4	-9.6	-10.2		
9:00	13.4	8.9	12.8	8.6	12.5	9.4	10.6	7.8	-8.7	-10.0		
10:00	13.4	9.5	13.0	9.5	12.8	9.9	10.7	7.4	-7.0	-8.6		
11:00	13.4	9.1	12.8	8.9	12.4	8.5	10.7	7.9	-5.7	-7.7		
Noon	13.4	9.6	12.9	9.5	12.8	9.8	11.2	8.8	-4.3	-6.2		
1:00 pm	13.4	9.5	12.9	9.4	13.0	10.6	11.6	8.5	-4.7	-6.3		
2:00	13.4	9.5	12.9	9.2	12.9	10.3	11.6	9.1	-4.7	-6.2		
3:00	13.4	10.0	13.3	9.6	13.4	11.2	11.8	8.9	-5.2	-6.6		
4:00	13.4	9.4	13.1	8.9	13.1	9.5	11.7	7.2	-5.0	-6.3		
5:00	13.3	9.0	12.8	8.9	12.4	8.4	10.1	7.8	-5.3	-6.2		
6:00	13.3	9.5	12.9	9.1	12.7	9.8	10.6	8.4	-5.7	-6.6		
7:00	13.3	9.9	12.9	9.5	12.9	10.1	10.7	8.4	-5.7	-6.7		
8:00	13.4	9.6	13.1	9.5	13.2	11.2	10.7	8.4	-5.7	-6.7		

Diurnal Temperature Measurements

COUNTY: Linfen City
PROVINCE: Shanxi
SETTLEMENT NAME: Da Yang Village
FAMILY/HOUSE NAME: Cui Mingxing

House No. 6 — Summer
Sep.4-5,1984
Degrees in Centigrade

Time	1 Dry	1 Wet	2 Dry	2 Wet	3 Dry	3 Wet	4 Dry	4 Wet	5 Dry	5 Wet	6 Dry	6 Wet
8:00 pm	20.4	18.9	21.6	18.9	22.8	18.9	22.5	19.0	22.3	18.4		
9:00	20.1	18.8	21.3	18.6	22.6	19.0	22.5	19.4	21.2	17.6		
10:00	20.1	18.6	21.2	18.6	22.7	19.4	22.4	19.2	20.1	17.7		
11:00	20.1	18.7	21.2	18.7	22.6	19.6	22.4	19.2	20.3	18.0		
Midnight	20.3	18.9	21.2	18.9	22.7	19.7	22.2	20.1	20.5	18.1		
1:00 am	20.1	19.0	21.3	18.9	22.4	19.6	22.6	19.8	20.0	18.7		
2:00	20.1	19.0	21.2	19.0	22.4	19.7	22.6	20.1	19.8	18.1		
3:00	20.1	18.9	21.2	18.8	22.4	20.1	22.6	19.9	19.6	18.7		
4:00	20.1	18.9	21.3	18.9	22.4	19.8	22.6	19.7	18.8	17.8		
5:00	20.1	18.9	21.2	18.9	22.3	19.8	22.4	19.4	17.0	16.3		
6:00	20.1	18.9	21.2	18.8	22.2	19.6	22.3	19.5	17.0	16.1		
7:00	20.0	18.9	21.2	18.7	21.8	19.0	22.2	19.0	18.2	17.1		
8:00	20.0	18.8	21.2	18.8	21.8	19.2	22.0	19.1	19.0	18.1		
9:00	20.1	18.9	21.2	18.9	22.2	20.1	21.8	19.5	20.1	18.1		
10:00	20.1	18.9	21.2	18.9	22.2	19.4	22.2	19.7	21.2	18.9		
11:00	20.1	18.9	21.2	18.8	22.1	19.3	22.2	20.1	20.6	18.8		
Noon	20.1	18.9	21.2	18.7	22.1	19.2	22.2	20.4	20.4	18.4		
1:00 pm	20.1	18.8	21.2	18.8	22.0	19.6	22.0	19.6	20.3	18.8		
2:00	20.1	18.8	21.2	18.9	22.0	19.6	22.0	19.0	20.1	17.6		
3:00	20.1	18.8	21.2	18.8	22.0	19.0	21.8	19.1	20.1	17.6		
4:00	20.1	18.8	21.1	18.9	21.8	18.9	21.7	18.3	19.2	17.3		
5:00	19.9	18.8	21.0	18.6	21.5	18.5	21.3	18.2	18.8	17.2		
6:00	20.0	18.6	21.0	18.6	21.7	19.0	21.4	19.0	18.6	17.3		
7:00	20.0	18.7	21.0	18.6	21.6	19.0	21.3	18.9	18.4	17.6		
8:00	20.0	18.9	21.1	19.0	21.8	19.6	21.8	19.8	18.2	17.4		

Relative Humidity Calculations

COUNTY: Linfen City
PROVINCE: Shanxi
SETTLEMENT NAME: Da Yang Village
FAMILY/HOUSE NAME: Cui Mingxing

House No. 6 — Summer

Time	SITES					
	1 RH%	2 RH%	3 RH%	4 RH%	5 RH%	6 RH%
8:00 pm	88.0	78.5	71.1	73.3	70.4	
9:00	89.2	78.4	72.6	76.5	71.7	
10:00	87.4	78.7	74.6	74.9	80.8	
11:00	88.8	79.6	76.9	75.3	80.9	
Midnight	88.8	81.7	77.0	83.3	80.5	
1:00 am	91.5	81.3	78.1	78.5	89.2	
2:00	91.1	82.6	78.5	80.6	86.0	
3:00	90.6	80.9	81.8	79.8	91.9	
4:00	90.1	81.3	79.7	78.1	91.3	
5:00	90.1	81.7	80.1	76.9	93.4	
6:00	90.1	80.9	79.6	78.0	91.4	
7:00	90.6	80.4	77.8	75.5	90.7	
8:00	89.7	81.3	79.4	77.5	91.8	
9:00	90.1	82.1	82.9	81.5	83.9	
10:00	90.1	82.1	78.4	80.0	82.1	
11:00	90.6	80.8	78.3	83.3	85.8	
Noon	90.6	80.0	77.5	85.9	83.5	
1:00 pm	89.7	81.3	81.2	81.2	87.9	
2:00	89.7	81.7	80.8	77.1	79.5	
3:00	89.7	81.3	76.7	79.0	79.5	
4:00	89.7	82.1	77.8	74.0	83.5	
5:00	90.1	80.3	76.0	75.5	86.1	
6:00	88.3	80.3	79.0	80.5	89.3	
7:00	89.2	80.8	79.3	81.3	92.6	
8:00	90.6	83.4	82.8	84.0	93.1	

Relative Humidity Calculations

COUNTY: Linfen City
PROVINCE: Shanxi
SETTLEMENT NAME: Da Yang Village
FAMILY/HOUSE NAME: Cui Mingxing

House No. 6 — Winter

Time	SITES 1 RH%	2 RH%	3 RH%	4 RH%	5 RH%	6 RH%
8:00 pm	67.8	63.3	71.2	65.8	78.5	
9:00	63.9	63.0	66.4	62.2	77.2	
10:00	67.4	65.1	67.5	65.4	81.7	
11:00	67.4	65.1	66.9	65.8	72.2	
Midnight	64.4	62.0	64.5	65.8	80.6	
1:00 am	64.4	60.9	65.4	66.7	77.4	
2:00	64.9	59.0	66.8	67.2	74.0	
3:00	62.9	59.9	66.8	65.5	70.0	
4:00	60.5	60.4	66.8	64.3	69.0	
5:00	60.5	61.4	66.8	63.2	69.0	
6:00	61.0	62.3	67.7	62.0	72.3	
7:00	60.5	62.3	67.7	63.5	70.2	
8:00	62.9	62.3	61.7	63.7	83.4	
9:00	58.0	59.3	69.1	70.6	68.7	
10:00	63.4	66.5	71.5	65.2	63.2	
11:00	59.5	62.3	61.8	70.7	57.6	
Noon	63.9	66.9	70.4	75.0	63.9	
1:00 pm	62.9	66.0	76.3	68.3	67.5	
2:00	63.4	63.9	74.2	74.7	70.8	
3:00	67.4	64.8	77.7	70.6	71.2	
4:00	62.0	59.7	65.6	55.3	71.5	
5:00	59.4	62.3	60.8	75.3	82.4	
6:00	63.8	63.4	71.4	76.8	82.0	
7:00	66.9	66.5	71.6	75.7	78.5	
8:00	63.9	65.6	80.1	75.2	78.5	

Diurnal Temperature Measurements

COUNTY: Qian Xian
PROVINCE: Shaanxi
SETTLEMENT NAME: Shima Dao Cun Village
FAMILY/HOUSE NAME: Bai Lesheng

House No. 7 — Winter
Dec. 7-8, 1984
Degrees in Centigrade

Time	1 Dry	1 Wet	2 Dry	2 Wet	3 Dry	3 Wet	4 Dry	4 Wet	5 Dry	5 Wet	6 Dry	6 Wet
8:00 pm	2.3	-0.9	4.7	1.5	4.8	0.6	0.9	-1.1	0.7	-1.3	0.8	-2.0
9:00	2.0	-0.9	4.7	1.9	4.5	1.0	0.7	-1.3	0.2	-1.3	0.8	-2.0
10:00	1.8	-0.4	4.5	1.7	3.9	1.0	0.3	-1.2	-0.2	-1.7	0.1	-2.6
11:00	2.0	-0.1	5.2	2.8	3.9	1.3	0.5	-1.1	-0.2	-2.0	0.3	-2.3
Midnight	2.0	-0.1	5.3	2.8	4.2	1.7	0.6	-1.1	-0.2	-2.0	0.1	-2.3
1:00 am	2.2	0.1	5.3	3.1	4.3	2.2	0.6	-1.2	-0.2	-2.1	0.1	-2.3
2:00	2.3	0.1	5.5	3.3	4.5	2.3	0.7	-1.1	-0.2	-2.1	-0.3	-2.3
3:00	2.4	-0.1	5.6	3.2	4.5	2.6	0.5	-0.5	-0.2	-2.1	-0.5	-2.3
4:00	2.4	0.3	5.6	3.2	4.6	2.4	0.4	-0.6	-0.3	-2.2	-0.5	-2.3
5:00	2.3	0.9	5.5	3.3	4.6	2.3	0.2	-0.8	-0.3	-2.1	-0.5	-2.0
6:00	2.3	0.5	5.6	3.3	4.6	2.5	0.4	-0.7	-0.6	-2.1	-0.3	-2.0
7:00	2.3	0.1	5.7	3.3	4.6	2.6	0.6	-0.6	-0.6	-1.2	-0.2	-2.2
8:00	2.2	0.5	5.4	2.6	4.8	2.7	0.6	-0.4	-0.2	-1.3	0.1	-1.7
9:00	2.2	0.6	5.7	3.1	4.8	2.8	0.6	-0.2	-0.1	-0.9	0.4	-1.1
10:00	2.3	0.7	5.2	2.2	4.9	2.3	0.7	0.1	0.1	-0.6	0.3	-0.7
11:00	2.3	1.2	4.8	2.4	4.6	2.7	0.6	-0.1	0.1	-0.7	0.1	-0.9
Noon	2.3	1.2	4.6	2.4	4.8	3.1	0.6	0.1	0.1	-0.9	-0.2	-1.1
1:00 pm	2.4	1.4	4.7	2.6	4.7	2.8	0.6	0.1	0.1	-0.7	-0.2	-0.9
2:00	2.3	1.2	4.8	2.3	5.1	3.3	0.2	-0.5	-0.3	-1.2	-0.7	-1.3
3:00	2.4	1.3	4.6	2.3	4.8	2.9	0.2	-0.2	-0.3	-1.2	-0.7	-1.5
4:00	2.3	0.7	4.6	2.4	4.6	2.7	0.4	-0.4	-0.3	-1.2	-0.5	-1.3
5:00	2.6	0.7	4.5	2.3	4.8	3.0	0.1	-0.6	-0.2	-1.1	-0.3	-1.2
6:00	2.4	0.7	4.5	2.7	4.7	2.8	0.1	-0.4	-0.2	-0.7	-0.3	-1.7
7:00	2.0	1.0	4.8	2.8	4.7	2.8	0.1	-0.4	-0.4	-1.2	0.1	-1.2
8:00	2.3	1.4	4.5	2.9	4.8	2.9	0.3	-0.3	-0.4	-1.3	-0.5	-1.5

Diurnal Temperature Measurements

COUNTY: Qian Xian
PROVINCE: Shaanxi
SETTLEMENT NAME: Shima Dao Cun Village
FAMILY/HOUSE NAME: Bai Lesheng

House No. 7 — Summer
Aug. 27-28, 1984
Degrees in Centigrade

Time	1 Dry	1 Wet	2 Dry	2 Wet	3 Dry	3 Wet	4 Dry	4 Wet	5 Dry	5 Wet	6 Dry	6 Wet
8:00 pm	20.9	19.2	20.1	18.9	20.1	19.3	19.0	18.7	18.4	18.1	19.5	18.1
9:00	20.7	19.2	20.2	18.5	20.1	19.0	18.3	18.1	18.1	17.9	18.7	17.8
10:00	21.3	18.9	20.2	18.5	20.2	18.8	17.9	17.9	17.8	17.8	17.8	17.8
11:00	20.7	18.9	20.1	18.4	19.6	18.7	17.4	17.4	17.3	17.3	17.3	17.3
Midnight	20.6	18.4	20.0	18.4	19.0	18.4	17.2	17.2	17.2	17.0	19.5	18.2
1:00 am	20.6	18.4	20.0	18.4	19.0	18.1	17.6	17.6	17.4	17.0	18.4	17.0
2:00	21.1	18.4	20.0	18.4	19.0	18.1	18.2	17.7	17.6	17.0	18.4	17.2
3:00	20.9	18.4	19.8	18.3	18.9	18.3	18.1	17.6	17.6	17.3	18.1	17.3
4:00	20.9	18.4	19.8	18.3	18.9	18.3	17.9	17.8	17.6	17.3	17.8	17.3
5:00	20.5	18.4	19.7	18.4	18.9	18.3	17.8	17.7	17.3	17.3	18.0	17.3
6:00	20.1	18.3	18.4	17.8	18.9	18.3	17.3	17.3	17.3	17.2	17.2	16.7
7:00	19.0	17.3	18.3	16.9	18.7	17.7	16.4	16.4	16.2	16.2	16.3	15.9
8:00	19.2	17.4	18.4	17.2	17.8	17.9	16.7	16.7	16.8	16.4	17.0	16.3
9:00	19.5	17.5	19.2	17.5	18.4	17.9	18.2	17.6	17.6	16.9	17.6	16.7
10:00	19.9	17.9	19.5	18.1	18.9	17.9	19.6	17.7	18.9	17.3	18.9	17.9
11:00	20.4	18.7	19.7	18.3	19.0	18.8	20.6	18.4	20.1	18.4	20.1	18.3
Noon	20.6	18.8	19.9	18.4	19.0	18.7	20.1	19.5	21.7	19.5	20.9	19.2
1:00 pm	20.9	18.9	20.0	18.4	19.1	18.4	21.1	19.0	20.1	18.9	20.3	18.5
2:00	21.1	19.5	20.1	18.4	19.2	18.7	22.3	19.6	20.2	19.0	20.1	18.7
3:00	21.2	19.5	20.3	19.0	19.4	19.0	22.3	20.1	21.4	19.5	22.3	20.1
4:00	21.3	20.0	20.6	19.4	19.6	19.0	22.8	19.9	21.6	19.7	21.2	19.6
5:00	21.3	19.6	20.6	19.4	19.6	19.0	21.3	19.5	20.7	19.5	20.9	19.5
6:00	21.2	19.5	20.4	18.9	19.6	19.0	20.7	19.4	19.7	18.9	19.8	18.4
7:00	21.2	19.6	20.2	18.7	19.4	19.0	19.9	18.9	18.9	18.4	19.0	18.1
8:00	20.9	19.2	20.2	18.9	19.6	19.2	19.4	18.7	18.7	18.3	18.9	18.3

Relative Humidity Calculations

COUNTY: Qian Xian
PROVINCE: Shaanxi
SETTLEMENT NAME: Shima Dao Cun Village
FAMILY/HOUSE NAME: Bai Lesheng

House No. 7 — Summer

Time	1 RH%	2 RH%	3 RH%	4 RH%	5 RH%	6 RH%
8:00 pm	86.3	90.2	92.9	97.0	97.5	88.2
9:00	88.0	85.7	91.1	98.5	98.5	92.7
10:00	80.5	86.1	88.4	100.0	100.0	100.0
11:00	85.4	86.1	91.9	100.0	100.0	100.0
Midnight	82.3	86.5	94.6	100.0	97.9	81.4
1:00 am	81.9	86.5	91.8	100.0	98.9	87.9
2:00	78.3	86.5	91.8	95.5	94.9	89.3
3:00	79.9	87.8	94.1	95.5	97.4	92.6
4:00	79.9	87.3	95.1	99.4	97.4	95.0
5:00	82.7	88.7	95.1	99.4	99.4	93.5
6:00	85.2	95.0	95.1	100.0	98.9	95.9
7:00	84.8	87.4	91.7	99.4	100.0	96.3
8:00	84.4	89.7	93.6	100.0	96.4	93.4
9:00	83.2	85.8	95.0	94.0	93.9	92.0
10:00	83.4	88.2	91.3	83.7	86.2	90.8
11:00	85.3	88.7	98.5	82.3	86.5	85.6
Noon	85.4	87.8	96.6	95.2	82.3	86.3
1:00 pm	84.2	87.0	94.2	83.4	90.6	84.8
2:00	87.3	86.5	95.6	78.8	90.2	88.3
3:00	86.4	88.9	97.1	82.5	84.4	82.5
4:00	89.5	90.3	95.6	77.9	84.4	86.4
5:00	86.5	88.9	95.2	85.6	90.3	88.5
6:00	86.0	88.0	95.2	89.4	95.2	88.7
7:00	86.9	87.5	96.1	91.5	95.1	91.8
8:00	85.9	89.7	97.1	94.2	96.5	94.1

Relative Humidity Calculations

COUNTY: Qian Xian
PROVINCE: Shaanxi
SETTLEMENT NAME: Shima Dao Cun Village
FAMILY/HOUSE NAME: Bai Lesheng

House No. 7   Winter

Time	SITES 1 RH%	2 RH%	3 RH%	4 RH%	5 RH%	6 RH%
8:00 pm	56.3	58.9	46.9	71.0	69.7	58.7
9:00	59.2	63.1	55.2	69.0	75.6	58.7
10:00	68.9	62.8	60.5	76.6	75.0	59.2
11:00	70.9	69.2	64.1	75.2	71.5	59.8
Midnight	70.9	68.6	67.3	74.5	71.5	62.6
1:00 am	71.1	71.4	71.7	72.1	69.0	62.6
2:00	68.5	71.6	70.5	73.9	69.0	67.9
3:00	67.0	69.0	74.1	84.6	69.0	70.2
4:00	69.5	69.0	70.6	84.4	69.6	70.2
5:00	79.6	71.6	69.2	84.2	71.3	73.9
6:00	74.0	70.4	72.1	81.9	74.5	73.9
7:00	69.8	69.8	73.5	81.3	89.8	68.1
8:00	75.5	64.0	72.3	83.9	81.1	71.8
9:00	77.1	67.1	74.4	87.3	85.6	76.8
10:00	76.5	61.7	65.5	90.1	90.1	84.2
11:00	84.4	68.7	74.2	89.9	88.3	83.9
Noon	83.6	70.6	77.9	91.7	83.9	85.5
1:00 pm	86.1	71.5	75.0	89.3	88.3	88.2
2:00	82.8	66.7	77.5	88.5	85.4	89.7
3:00	82.9	69.2	74.4	93.0	85.4	87.9
4:00	76.5	71.4	74.9	86.2	85.4	86.2
5:00	72.1	70.5	76.5	88.4	86.4	86.3
6:00	75.0	75.6	75.7	92.7	90.9	77.4
7:00	85.1	73.7	75.0	92.7	87.1	79.6
8:00	87.6	77.7	75.1	90.4	85.3	83.5

Diurnal Temperature Measurements

COUNTY: Qing Yang Region
PROVINCE: Gansu
SETTLEMENT NAME: Xi Feng Zhen Town
FAMILY/HOUSE NAME: Xing Xigeng

House No. 8   Winter
Dec.5-6,1984
Degrees in Centigrade

Time	1 Dry	1 Wet	2 Dry	2 Wet	3 Dry	3 Wet	4 Dry	4 Wet	5 Dry	5 Wet	6 Dry	6 Wet
8:00 pm	8.5	4.9	9.9	5.1	5.8	2.2	-0.4	-2.4	-2.7	-3.7	-2.8	-4.0
9:00	7.3	4.2	10.0	5.3	5.7	1.8	-0.5	-4.4	-3.2	-5.0	-3.4	-5.2
10:00	7.9	5.2	11.2	7.2	6.3	2.7	-0.5	-4.7	-5.0	-5.9	-5.2	-6.7
11:00	8.4	5.6	11.4	7.7	6.7	3.1	-1.6	-5.0	-5.2	-5.9	-5.7	-7.3
Midnight	8.4	5.1	11.4	7.3	6.6	2.8	-1.6	-5.5	-7.1	-8.1	-6.7	-7.7
1:00 am	8.4	6.0	11.4	7.6	6.6	3.3	-2.2	-5.6	-5.7	-5.9	-5.7	-8.1
2:00	8.5	6.0	11.3	7.4	6.6	3.0	-2.2	-5.9	-6.7	-8.7	-6.2	-8.5
3:00	8.3	5.8	11.3	7.3	6.4	2.7	-2.5	-5.9	-6.7	-9.0	-6.6	-7.7
4:00	8.4	5.8	11.2	7.0	6.4	2.8	-2.7	-6.1	-6.6	-9.0	-7.0	-8.1
5:00	8.4	5.7	11.2	6.8	6.6	3.0	-2.8	-6.3	-6.1	-9.3	-7.2	-8.5
6:00	8.1	5.7	11.2	6.7	6.6	3.0	-3.2	-6.3	-6.7	-9.0	-7.2	-8.7
7:00	8.1	5.6	10.7	6.3	6.7	3.1	-3.4	-6.4	-7.0	-8.7	-7.0	-9.0
8:00	7.9	5.1	11.7	7.7	6.4	2.3	-6.6	-5.5	-5.7	-7.0	-5.7	-9.0
9:00	7.7	3.4	11.6	6.2	6.4	1.8	-2.8	-5.0	-5.2	-7.0	-5.0	-7.7
10:00	7.3	3.7	11.3	5.7	6.4	2.3	-2.7	-4.7	-3.2	-5.0	-3.6	-7.3
11:00	7.4	3.4	11.2	6.7	6.5	2.7	-2.1	-3.9	-2.0	-5.9	-2.0	-5.9
Noon	7.3	4.2	12.0	7.9	6.2	2.9	-1.3	-2.5	-2.0	-2.6	1.4	-3.3
1:00 pm	7.8	4.6	12.9	8.6	6.2	2.3	-0.8	-2.2	-2.3	-2.5	1.4	-2.6
2:00	7.6	4.5	11.4	7.4	6.2	3.1	-0.3	-1.6	-2.9	-3.0	1.8	-1.3
3:00	7.7	4.2	10.6	4.5	6.7	3.5	-0.2	-1.6	-2.3	-3.0	2.4	-1.2
4:00	7.8	4.1	10.3	3.9	6.7	2.8	0.1	-1.6	-1.6	-1.8	1.6	-0.9
5:00	8.2	4.5	9.5	3.8	6.6	2.8	0.1	-1.5	-0.2	-1.7	0.7	-2.2
6:00	8.0	4.5	10.1	6.2	6.4	2.9	0.1	-1.4	0.1	-1.7	1.4	-2.3
7:00	7.9	5.2	10.6	6.7	6.4	3.3	0.1	-1.1	0.1	-1.7	0.1	-2.5
8:00	8.8	5.1	10.7	7.3	6.3	2.6	0.1	-1.1	0.1	-1.7	1.4	-2.6

Diurnal Temperature Measurements

COUNTY: Qing Yang Region
PROVINCE: Gansu
SETTLEMENT NAME: Xi Feng Zhen Town
FAMILY/HOUSE NAME: Xing Xigeng

House No. 8   Summer
Jul31-Aug.1,1984
Degrees in Centigrade

Time	1 Dry	1 Wet	2 Dry	2 Wet	3 Dry	3 Wet	4 Dry	4 Wet	5 Dry	5 Wet	6 Dry	6 Wet
8:00 pm	21.6	20.6	23.7	21.4	21.9	21.3	20.9	20.1	21.2	21.2	23.1	21.2
9:00	21.3	20.2	23.4	21.3	21.3	20.7	20.7	20.1	21.6	20.7	22.8	20.9
10:00	21.8	20.6	23.4	21.7	21.2	20.7	20.7	20.1	21.2	20.6	23.1	20.7
11:00	21.5	20.5	23.4	21.4	21.1	20.7	20.5	19.6	21.1	20.6	22.3	20.6
Midnight	21.2	20.2	23.1	21.3	20.9	20.6	20.2	19.6	21.1	20.3	21.7	20.4
1:00 am	21.2	20.1	22.9	21.2	20.7	20.3	20.2	19.5	20.1	20.1	21.8	20.2
2:00	21.1	20.0	22.8	21.2	20.6	20.2	20.2	19.4	20.1	19.9	22.3	20.1
3:00	21.1	19.9	22.7	21.1	20.6	20.1	20.1	19.3	20.1	19.8	21.8	20.2
4:00	21.1	19.9	22.6	20.7	20.3	19.9	20.0	19.0	19.9	19.6	21.6	20.3
5:00	20.8	19.6	22.4	20.7	20.2	19.6	19.9	19.0	19.2	19.0	20.9	19.3
6:00	20.6	19.8	22.3	20.5	20.1	19.6	19.8	18.9	19.4	19.2	21.7	19.7
7:00	20.6	19.5	21.9	20.1	20.1	19.5	19.8	19.2	20.1	19.5	23.9	20.1
8:00	20.8	19.6	22.2	20.6	20.2	19.9	19.9	19.5	22.3	20.1	22.8	20.3
9:00	21.1	19.8	22.9	21.2	20.6	19.9	20.1	19.6	23.1	20.6	22.8	20.3
10:00	22.2	20.7	22.9	21.2	20.4	20.2	20.1	19.7	21.7	20.1	23.1	20.6
11:00	22.7	20.8	22.8	21.3	20.6	20.5	20.1	19.6	21.3	20.6	23.1	21.1
Noon	21.7	20.3	23.4	22.2	20.8	20.9	20.3	19.9	25.3	21.8	24.8	21.3
1:00 pm	21.5	20.3	23.4	21.7	22.3	21.2	20.3	19.8	26.6	22.3	25.6	21.8
2:00	21.7	20.6	23.4	21.7	21.3	21.2	20.4	19.9	29.6	23.1	26.2	22.6
3:00	21.8	21.1	23.7	22.3	21.4	21.2	20.7	20.2	26.2	21.7	26.1	22.0
4:00	21.8	20.3	23.6	21.4	21.6	20.7	20.7	20.1	26.8	21.7	25.9	21.7
5:00	21.7	20.1	23.4	21.2	21.3	20.6	20.6	19.6	24.0	21.2	24.9	20.9
6:00	21.4	20.3	23.5	21.4	21.2	20.5	20.5	19.9	23.7	20.6	24.0	20.6
7:00	21.2	19.8	23.9	21.7	20.9	20.3	20.3	19.6	21.6	20.1	23.4	20.1
8:00	21.3	20.1	23.7	21.4	20.9	20.6	20.9	20.0	20.1	19.8	22.3	20.2

Relative Humidity Calculations

COUNTY: Qing Yang Region
PROVINCE: Gansu
SETTLEMENT NAME: Xi Feng Zhen Town
FAMILY/HOUSE NAME: Xing Xigeng

House No. 8   Summer

Time	SITES 1 RH%	2 RH%	3 RH%	4 RH%	5 RH%	6 RH%
8:00 pm	91.4	82.7	95.0	93.5	100.0	85.3
9:00	90.9	83.8	94.9	94.8	92.3	84.8
10:00	90.1	87.1	95.8	95.3	95.3	81.6
11:00	91.8	85.0	97.2	92.5	95.8	86.8
Midnight	91.7	85.7	97.6	94.8	93.5	89.6
1:00 am	90.8	86.1	96.2	94.3	100.0	86.6
2:00	90.8	86.9	96.2	93.4	99.0	83.0
3:00	90.4	87.3	95.3	93.4	97.6	87.0
4:00	90.4	85.6	96.7	92.0	97.1	89.6
5:00	90.3	86.0	94.8	92.4	98.5	87.2
6:00	93.0	85.9	95.2	92.4	98.5	83.6
7:00	90.7	85.4	94.8	95.2	95.2	71.5
8:00	89.9	87.2	96.6	96.6	82.5	80.7
9:00	89.5	86.1	94.4	95.2	80.8	80.7
10:00	88.0	86.1	97.6	96.6	86.6	80.8
11:00	85.2	87.8	99.5	95.7	94.0	84.5
Noon	88.7	90.4	100.0	96.7	74.8	74.6
1:00 pm	90.5	87.1	91.1	95.3	70.1	73.1
2:00	90.9	87.1	99.5	95.7	59.6	74.5
3:00	94.1	89.2	97.6	95.8	69.1	71.2
4:00	88.3	83.0	92.7	94.8	65.7	70.8
5:00	87.0	83.0	94.4	92.1	78.8	70.9
6:00	90.9	83.8	94.0	94.8	77.1	74.9
7:00	88.6	83.1	94.9	93.9	87.4	75.0
8:00	90.0	83.0	97.6	92.6	97.6	83.4

Relative Humidity Calculations

COUNTY: Qing Yang Region
PROVINCE: Gansu
SETTLEMENT NAME: Xi Feng Zhen Town
FAMILY/HOUSE NAME: Xing Xigeng

House No. 8   Winter

Time	\multicolumn{6}{c}{SITES}					
	1 RH%	2 RH%	3 RH%	4 RH%	5 RH%	6 RH%
8:00 pm	60.1	49.4	56.0	67.8	81.6	79.5
9:00	64.0	51.2	52.6	39.0	67.0	67.6
10:00	67.8	59.3	56.3	35.0	79.4	68.9
11:00	68.3	62.2	56.9	44.3	83.7	65.6
Midnight	62.4	58.1	55.4	35.6	76.7	77.2
1:00 am	73.2	60.7	60.5	41.0	95.1	49.6
2:00	71.5	60.0	57.3	36.3	55.1	50.7
3:00	71.3	59.0	55.2	39.2	50.3	73.7
4:00	70.2	57.2	56.5	41.1	47.2	73.1
5:00	68.9	56.1	57.3	38.9	31.4	70.2
6:00	71.7	55.1	56.7	43.6	50.3	63.9
7:00	71.1	54.4	56.9	44.8	59.4	54.5
8:00	66.6	60.0	50.2	127.2	73.8	32.9
9:00	51.3	47.5	45.3	61.7	63.3	44.3
10:00	57.3	45.1	50.8	64.9	67.0	30.7
11:00	54.3	54.6	54.7	67.9	68.3	32.8
Noon	63.4	59.8	60.5	81.0	68.5	33.6
1:00 pm	63.4	58.4	53.5	77.0	69.0	43.4
2:00	64.4	59.1	61.8	79.2	69.6	55.8
3:00	59.7	38.8	61.3	77.5	72.3	51.4
4:00	56.8	35.7	54.3	73.5	73.7	64.4
5:00	59.1	40.7	55.4	75.2	75.0	57.5
6:00	60.6	58.9	57.1	77.0	71.8	47.0
7:00	67.8	59.5	61.6	81.4	71.8	60.9
8:00	58.3	64.6	54.9	81.4	71.8	44.0

Diurnal Temperature Measurements

COUNTY: Xing Yang
PROVINCE: Henan
SETTLEMENT NAME: Bei Tai
FAMILY/HOUSE NAME: Tian Lu

House No. 9   Winter
Dec.24-25,1984
Degrees in Centigrade

Time	1 Dry	1 Wet	2 Dry	2 Wet	3 Dry	3 Wet	4 Dry	4 Wet	5 Dry	5 Wet	6 Dry	6 Wet
8:00 pm	5.5	1.3	6.0	2.8	1.7	0.1	-5.0	-6.2	-5.2	-6.1		
9:00	6.7	2.8	5.6	1.3	1.3	-0.8	-5.6	-6.7	-5.3	-5.9		
10:00	6.7	2.8	5.2	1.0	1.3	-0.9	-5.3	-6.2	-5.6	-6.6		
11:00	5.9	2.9	5.2	1.2	1.2	-0.9	-5.6	-6.7	-5.7	-7.0		
Midnight	5.3	3.3	5.2	1.6	1.2	-0.9	-5.9	-7.0	-6.2	-7.3		
1:00 am	4.9	2.9	5.2	1.3	1.2	-1.0	-6.1	-7.1	-5.9	-7.6		
2:00	4.7	2.5	5.3	0.7	1.2	-1.1	-5.9	-7.6	-6.6	-8.0		
3:00	4.8	2.7	5.2	1.0	1.2	-1.1	-5.8	-7.1	-6.2	-7.6		
4:00	4.9	2.9	5.1	1.2	1.1	-1.1	-5.3	-6.6	-5.7	-7.0		
5:00	4.9	2.9	4.8	1.3	1.2	-1.1	-5.3	-7.2	-6.2	-7.2		
6:00	5.0	2.9	4.6	1.4	1.2	-1.1	-5.6	-7.0	-6.6	-7.3		
7:00	4.9	2.9	4.7	1.0	1.2	-0.7	-4.8	-6.7	-4.7	-5.9		
8:00	4.9	2.9	4.7	0.6	1.3	-0.5	-4.6	-6.6	-4.6	-6.6		
9:00	4.4	1.4	5.2	1.8	1.7	-0.5	-3.2	-5.0	-5.7	-5.3		
10:00	5.2	1.0	5.6	2.2	1.9	0.6	-1.6	-3.2	-1.2	-4.2		
11:00	4.5	1.7	5.6	2.3	2.4	0.6	-0.7	-1.7	-1.1	-2.7		
Noon	4.6	1.8	6.1	3.1	2.8	1.2	0.1	-1.2	0.1	-2.6		
1:00 pm	4.7	1.3	6.3	3.4	3.1	1.1	0.7	-1.1	-1.3	-0.6		
2:00	5.1	2.7	6.6	3.7	3.4	1.7	1.3	-1.2	1.8	-1.2		
3:00	5.1	2.7	6.4	3.8	3.4	1.4	-1.2	-1.2	1.8	-2.2		
4:00	5.1	1.3	6.7	3.1	3.2	1.3	-0.6	-2.6	1.0	-2.9		
5:00	4.7	2.3	6.7	3.6	3.1	1.0	0.1	-3.2	1.4	-3.3		
6:00	4.7	2.3	6.4	4.1	2.9	1.2	-0.3	-3.3	0.4	-3.7		
7:00	4.7	2.2	7.3	5.1	3.0	1.6	-0.7	-3.7	-0.7	-4.0		
8:00	5.2	2.9	7.3	4.9	2.8	1.4	-1.2	-4.2	-1.1	-4.0		

Diurnal Temperature Measurements

COUNTY: Xing Yang
PROVINCE: Henan
SETTLEMENT NAME: Bei Tai
FAMILY/HOUSE NAME: Tian Lu

House No. 9   Summer
Aug.4-5,1984
Degrees in Centigrade

Time	1 Dry	1 Wet	2 Dry	2 Wet	3 Dry	3 Wet	4 Dry	4 Wet	5 Dry	5 Wet	6 Dry	6 Wet
8:00 pm	26.4	24.6	25.8	23.9	23.9	23.4	25.9	24.9	27.3	23.8		
9:00	26.9	24.7	25.6	23.9	23.6	22.6	25.3	24.0	27.3	23.4		
10:00	26.5	24.0	25.2	23.7	23.4	22.8	24.5	23.4	26.2	23.4		
11:00	26.5	24.6	24.7	24.0	23.3	22.9	24.5	23.7	26.2	23.7		
Midnight	26.4	24.1	24.7	24.0	23.2	22.8	24.4	23.6	26.2	22.9		
1:00 am	26.2	24.0	24.5	23.9	22.9	22.4	24.0	23.4	25.9	23.1		
2:00	25.7	23.9	24.5	23.5	22.8	22.6	24.7	23.4	26.7	22.3		
3:00	25.6	23.8	24.6	22.9	22.8	22.3	25.1	22.6	26.7	22.6		
4:00	25.6	23.4	24.5	22.7	22.7	22.2	25.1	23.4	25.9	22.3		
5:00	25.6	23.7	24.2	23.4	22.7	22.3	24.6	23.4	25.3	23.1		
6:00	25.6	24.0	24.3	23.4	22.6	22.3	24.6	23.5	25.6	23.1		
7:00	25.6	24.1	24.5	23.9	22.6	22.2	25.8	23.9	26.4	23.4		
8:00	26.2	24.0	24.6	23.9	22.6	22.3	27.7	25.2	27.6	23.3		
9:00	25.6	23.5	25.2	23.4	23.4	22.7	32.5	25.1	31.2	25.1		
10:00	25.7	24.0	24.9	23.8	23.9	23.2	37.0	25.6	33.4	25.6		
11:00	25.7	24.6	25.6	23.9	23.2	22.9	43.7	26.7	32.3	25.5		
Noon	26.6	24.2	27.0	24.4	23.2	22.8	39.6	26.2	32.3	25.4		
1:00 pm	27.4	24.9	27.1	24.9	23.3	23.1	34.1	25.7	34.4	26.6		
2:00	26.9	25.1	28.7	25.2	23.3	22.8	34.1	26.3	33.4	26.7		
3:00	27.0	25.7	27.0	25.6	23.4	23.2	33.6	26.8	35.2	28.4		
4:00	27.9	26.1	26.5	25.2	23.9	23.4	32.4	26.4	34.8	27.8		
5:00	27.4	25.7	26.3	25.1	23.4	22.8	31.9	26.7	33.7	27.9		
6:00	27.3	25.6	26.2	25.1	23.4	23.1	30.1	26.8	32.2	26.8		
7:00	27.3	25.7	26.5	25.7	23.4	23.0	28.7	26.6	30.1	27.6		
8:00	27.7	26.3	26.4	25.6	23.9	23.8	27.9	26.7	29.5	26.8		

Relative Humidity Calculations

COUNTY: Xing Yang
PROVINCE: Henan
SETTLEMENT NAME: Bei Tai
FAMILY/HOUSE NAME: Tian Lu

House No. 9   Summer

Time	1 RH%	2 RH%	3 RH%	4 RH%	5 RH%	6 RH%
8:00 pm	86.6	86.1	95.6	92.9	76.1	
9:00	84.1	87.6	92.1	90.8	73.3	
10:00	82.1	88.7	95.1	91.5	79.7	
11:00	85.9	95.3	96.9	93.5	82.0	
Midnight	83.6	95.3	96.9	93.5	77.1	
1:00 am	83.9	95.7	96.0	95.2	80.0	
2:00	86.9	92.3	97.7	90.3	69.5	
3:00	86.8	87.8	95.5	81.6	71.2	
4:00	84.1	86.5	95.5	87.5	74.4	
5:00	86.0	93.5	96.4	91.1	83.6	
6:00	88.0	93.1	97.7	91.9	81.8	
7:00	88.8	95.7	96.8	86.1	78.3	
8:00	84.3	94.8	97.7	82.1	71.0	
9:00	84.5	86.7	94.3	56.7	62.5	
10:00	87.6	91.5	93.9	41.9	55.4	
11:00	91.3	87.6	97.7	28.6	59.4	
Noon	82.5	81.5	96.9	36.7	59.1	
1:00 pm	82.0	84.5	98.2	52.9	56.0	
2:00	87.2	76.0	96.0	56.1	60.9	
3:00	90.7	89.9	98.6	60.2	61.4	
4:00	87.0	89.8	97.8	63.8	60.4	
5:00	87.7	90.6	95.6	67.7	65.7	
6:00	88.0	91.7	97.7	78.0	66.7	
7:00	88.4	93.4	96.9	85.7	83.2	
8:00	89.6	94.2	99.1	91.6	82.0	

Relative Humidity Calculations

COUNTY: Xing Yang
PROVINCE: Henan
SETTLEMENT NAME: Bei Tai
FAMILY/HOUSE NAME: Tian Lu

House No. 9 Winter

Time	SITES 1 RH%	2 RH%	3 RH%	4 RH%	5 RH%	6 RH%
8:00 pm	47.8	60.4	76.6	74.9	80.3	
9:00	53.2	47.9	69.5	75.2	88.2	
10:00	53.7	48.4	67.7	82.4	78.7	
11:00	62.8	50.4	68.4	77.4	73.8	
Midnight	73.5	55.0	68.4	78.2	74.3	
1:00 am	73.1	51.1	67.6	78.0	63.0	
2:00	71.4	42.9	67.5	64.2	68.8	
3:00	73.0	47.7	67.5	73.6	69.5	
4:00	73.1	50.8	68.2	74.3	73.8	
5:00	73.8	54.4	67.5	61.9	77.9	
6:00	73.2	58.8	67.5	70.6	82.4	
7:00	73.8	53.4	73.2	61.8	76.3	
8:00	73.8	47.5	74.2	60.2	60.2	
9:00	60.6	57.7	69.3	67.0	54.6	
10:00	48.4	57.6	80.9	72.4	50.7	
11:00	62.2	58.4	73.4	83.3	72.1	
Noon	63.6	63.0	76.2	80.4	59.2	
1:00 pm	56.1	64.6	71.9	73.9	58.7	
2:00	69.1	64.3	75.9	64.5	58.1	
3:00	69.1	68.0	72.2	65.7	44.5	
4:00	52.1	56.8	72.7	67.5	42.2	
5:00	68.0	62.6	71.1	50.2	33.6	
6:00	68.7	70.6	74.7	52.6	38.0	
7:00	67.9	73.4	79.4	51.8	48.4	
8:00	71.3	72.1	80.0	50.7	52.7	

Diurnal Temperature Measurements

COUNTY: Gong Xian Township
PROVINCE: Henan
SETTLEMENT NAME: Xi Cun Village
FAMILY/HOUSE NAME: Yin Xin Yin

House No. 10 Winter
Dec.28-29,1984
Degrees in Centigrade

Time	1 Dry	1 Wet	2 Dry	2 Wet	3 Dry	3 Wet	4 Dry	4 Wet	5 Dry	5 Wet	6 Dry	6 Wet
8:00 pm	4.9	-0.6	4.5	0.1	1.8	-2.2	-2.2	-5.0	-3.2	-5.7		
9:00	4.0	-0.9	4.4	-0.7	1.8	-2.2	-2.5	-5.0	-2.7	-5.7		
10:00	4.1	0.6	4.3	0.0	1.6	-2.7	-2.6	-4.3	-2.7	-5.7		
11:00	4.0	0.4	4.2	-0.6	1.6	-2.4	-2.8	-5.5	-4.0	-5.8		
Midnight	3.9	0.3	4.1	-0.9	1.6	-2.2	-3.2	-5.9	-4.0	-5.9		
1:00 am	3.8	0.2	4.1	-0.7	1.5	-2.3	-3.2	-5.7	-4.0	-6.1		
2:00	3.7	0.1	4.0	-0.6	1.4	-2.5	-3.2	-5.7	-4.0	-5.9		
3:00	3.5	0.1	3.9	-0.8	1.4	-2.6	-3.7	-5.7	-4.6	-6.6		
4:00	3.4	0.1	3.9	-1.1	1.4	-2.6	-4.0	-5.7	-5.0	-6.7		
5:00	3.4	0.1	3.9	-1.1	1.4	-2.6	-4.3	-5.9	-5.3	-7.0		
6:00	3.4	0.1	3.8	-1.2	1.3	-2.6	-3.7	-5.7	-5.0	-7.0		
7:00	3.4	0.0	3.7	-1.3	1.3	-2.6	-3.3	-5.7	-4.2	-6.7		
8:00	3.4	-0.8	3.7	-0.1	1.4	-2.3	-4.0	-6.1	-5.3	-7.0		
9:00	3.9	-0.3	3.9	0.1	1.6	-2.1	-2.3	-4.7	-2.3	-5.6		
10:00	3.9	0.6	3.9	0.0	1.6	-1.9	-1.2	-4.7	-1.3	-4.0		
11:00	3.8	0.1	4.4	0.0	1.7	-1.9	-0.8	-4.5	-1.2	-3.3		
Noon	3.0	-1.5	4.7	1.3	1.4	-2.2	0.4	-2.3	0.1	-2.5		
1:00 pm	3.4	-1.1	5.6	2.2	1.9	-1.6	1.4	-0.7	1.3	-1.7		
2:00	3.5	-1.1	5.6	1.8	2.2	-1.6	2.0	-0.7	1.7	-2.3		
3:00	3.9	-0.1	5.4	1.7	2.3	-1.1	2.0	-2.7	3.3	0.1		
4:00	3.9	-0.5	5.3	1.7	2.3	-1.1	1.4	-3.2	1.6	-3.3		
5:00	4.0	-0.5	5.5	1.8	2.4	-0.7	1.8	-2.7	2.0	-2.0		
6:00	3.3	-1.1	5.1	1.2	2.2	-1.6	-0.2	-4.2	-0.7	-3.2		
7:00	2.9	-1.1	5.1	1.0	1.7	-1.9	-0.2	-4.0	-1.1	-4.2		
8:00	2.8	-1.2	4.8	0.6	1.6	-1.9	-0.3	-4.2	-1.1	-4.3		

Diurnal Temperature Measurements

COUNTY: Gong Xian Township
PROVINCE: Henan
SETTLEMENT NAME: Xi Cun Village
FAMILY/HOUSE NAME: Yin Xin Yin

House No. 10 Summer
Aug.7-8,1984
Degrees in Centigrade

Time	1 Dry	1 Wet	2 Dry	2 Wet	3 Dry	3 Wet	4 Dry	4 Wet	5 Dry	5 Wet	6 Dry	6 Wet
8:00 pm	25.7	23.9	25.1	23.4	26.8	24.6	25.9	23.9	25.2	23.7		
9:00	25.6	24.0	25.1	23.1	26.8	25.1	25.3	23.9	24.5	23.8		
10:00	25.6	23.9	25.0	23.4	26.4	24.0	24.9	23.8	24.5	23.1		
11:00	25.6	24.2	25.0	23.1	26.6	24.6	24.8	23.9	23.9	23.3		
Midnight	25.6	24.2	24.9	23.4	26.6	24.6	24.3	23.7	23.4	23.1		
1:00 am	25.6	23.9	24.8	23.4	26.4	24.4	24.2	23.4	23.3	22.8		
2:00	25.3	23.7	24.7	23.1	26.2	24.0	23.6	22.8	22.8	22.3		
3:00	25.2	23.7	24.5	23.0	26.1	23.9	23.6	22.8	23.0	22.3		
4:00	25.1	23.4	24.5	23.1	26.1	23.9	23.7	22.8	22.8	22.3		
5:00	25.1	23.3	24.5	22.8	25.9	24.0	23.9	23.4	23.4	22.6		
6:00	25.1	23.7	24.5	22.8	25.9	24.0	24.2	23.4	23.4	22.8		
7:00	25.6	23.9	24.4	22.8	26.0	24.2	24.9	23.5	25.9	23.7		
8:00	25.3	23.9	24.5	22.8	26.2	24.5	25.7	24.2	25.9	24.4		
9:00	25.3	24.0	24.7	23.1	26.3	24.6	28.5	25.1	32.3	26.7		
10:00	25.3	23.9	24.8	23.5	26.5	24.9	28.9	25.6	28.7	25.9		
11:00	25.6	24.6	25.1	23.9	26.8	25.9	33.4	26.2	36.4	28.9		
Noon	25.7	24.6	25.5	24.0	27.0	25.5	32.2	25.5	31.7	26.2		
1:00 pm	25.8	24.4	25.6	23.0	28.3	26.2	32.2	25.7	33.9	27.8		
2:00	25.9	24.4	25.6	23.9	28.4	26.6	31.7	25.5	34.5	28.2		
3:00	26.2	24.1	25.6	23.7	28.2	25.6	31.7	24.7	33.4	26.8		
4:00	26.3	24.6	25.7	23.9	27.9	25.6	31.2	25.2	32.7	27.3		
5:00	26.4	24.6	25.7	23.9	28.0	26.2	30.6	25.6	31.4	26.1		
6:00	26.2	24.5	25.6	23.9	28.1	25.6	29.2	25.1	30.6	25.6		
7:00	26.2	24.5	25.6	23.9	27.8	25.6	28.7	25.1	29.6	25.2		
8:00	26.1	24.5	25.6	23.9	27.4	25.4	28.3	24.8	28.4	25.6		

Relative Humidity Calculations

COUNTY: Gong Xian Township
PROVINCE: Henan
SETTLEMENT NAME: Xi Cun Village
FAMILY/HOUSE NAME: Yin Xin Yin

House No. 10 Summer

Time	SITES 1 RH%	2 RH%	3 RH%	4 RH%	5 RH%	6 RH%
8:00 pm	87.2	87.5	84.1	85.7	88.7	
9:00	88.0	85.5	87.5	89.6	94.8	
10:00	88.0	87.9	83.2	91.5	89.4	
11:00	89.2	85.9	85.5	93.6	94.8	
Midnight	89.2	88.3	85.5	94.8	97.7	
1:00 am	88.0	89.5	85.1	93.9	96.4	
2:00	87.9	88.2	84.3	93.9	95.5	
3:00	88.7	88.6	84.2	94.3	94.2	
4:00	87.1	89.4	84.2	93.4	95.5	
5:00	86.7	87.4	85.7	96.0	93.4	
6:00	89.5	87.4	86.5	93.9	95.6	
7:00	87.6	88.1	86.9	89.1	83.4	
8:00	89.6	87.4	87.8	89.2	88.9	
9:00	90.0	87.8	87.8	76.7	66.1	
10:00	89.6	89.9	88.6	77.9	81.0	
11:00	92.5	91.2	93.4	58.4	58.7	
Noon	91.3	88.8	89.1	60.1	65.8	
1:00 pm	89.7	88.0	85.6	61.0	64.1	
2:00	88.9	87.6	87.5	62.5	63.5	
3:00	85.0	85.7	82.3	57.8	61.5	
4:00	87.8	86.9	83.7	63.1	67.2	
5:00	86.7	86.9	87.4	68.3	66.9	
6:00	87.8	87.6	82.6	72.5	68.3	
7:00	87.8	87.6	84.7	75.7	71.3	
8:00	88.5	87.6	86.1	75.9	80.9	

Relative Humidity Calculations

COUNTY: Gong Xian Township
PROVINCE: Henan
SETTLEMENT NAME: Xi Cun Village
FAMILY/HOUSE NAME: Yin Xin Yin

House No. 10   Winter

Time	SITES					
	1	2	3	4	5	6
	RH%	RH%	RH%	RH%	RH%	RH%
8:00 pm	35.0	46.5	45.1	51.9	53.2	
9:00	39.1	37.9	45.1	55.9	46.7	
10:00	53.1	47.4	39.4	68.1	46.7	
11:00	52.9	42.3	43.9	53.1	63.6	
Midnight	52.1	39.3	46.9	49.3	61.5	
1:00 am	52.5	42.0	45.8	53.2	58.4	
2:00	52.2	43.8	45.5	54.1	62.6	
3:00	55.7	41.6	43.4	61.1	60.2	
4:00	56.8	38.3	43.2	65.7	63.8	
5:00	56.8	38.3	43.2	68.2	66.4	
6:00	57.4	37.8	45.1	62.0	59.4	
7:00	56.7	37.5	45.1	55.8	50.7	
8:00	46.3	52.6	47.7	59.5	66.4	
9:00	47.5	52.3	48.4	58.0	44.1	
10:00	55.6	51.1	50.6	42.8	56.5	
11:00	50.5	46.4	49.4	42.2	63.9	
Noon	41.6	56.8	49.3	58.4	60.9	
1:00 pm	42.2	57.7	50.8	68.9	56.0	
2:00	42.4	52.5	48.4	61.6	44.0	
3:00	49.7	53.5	54.6	35.1	58.6	
4:00	45.0	54.6	54.0	35.2	32.0	
5:00	44.5	54.2	58.0	37.2	45.2	
6:00	44.5	51.5	47.9	38.4	60.3	
7:00	48.3	49.5	49.4	41.6	49.3	
8:00	48.0	46.9	50.6	40.3	46.7	

Diurnal Temperature Measurements

COUNTY: Gong Xian County
PROVINCE: Henan
SETTLEMENT NAME: Gong Xian Town
FAMILY/HOUSE NAME: Li Songbin

House No. 11   Winter
Dec.26-27,1984
Degrees in Centigrade

Time	SITES											
	1		2		3		4		5		6	
	Dry	Wet	Dry	Wet	Dry	Wet	Dry	Wet	Dry	Wet	Dry	Wet
8:00 pm	11.3	7.5	11.2	8.4	10.4	4.8	4.1	1.0	-2.9	-4.7	-2.3	-4.7
9:00	10.7	6.7	10.7	7.3	8.9	5.7	3.9	1.1	-3.2	-5.0	-2.9	-4.7
10:00	10.3	6.6	10.1	6.7	8.9	6.7	3.7	1.3	-3.7	-5.3	-3.2	-5.0
11:00	10.3	6.5	10.1	6.7	8.9	6.3	3.7	0.7	-4.0	-5.3	-3.4	-5.0
Midnight	10.3	6.4	9.9	7.0	8.8	6.3	3.6	0.7	-4.0	-5.3	-2.7	-5.3
1:00 am	10.2	6.3	9.7	7.2	8.7	6.2	3.5	0.6	-4.0	-5.6	-4.2	-5.6
2:00	10.2	6.3	9.9	7.0	8.6	6.2	3.5	0.5	-4.0	-5.6	-4.2	-5.7
3:00	10.1	6.3	9.5	7.0	8.4	6.1	3.4	0.4	-4.0	-5.7	-4.2	-5.9
4:00	10.1	6.2	9.5	7.0	8.4	6.1	3.4	0.8	-4.2	-6.6	-4.5	-6.2
5:00	10.1	6.1	9.5	7.0	8.4	6.1	3.3	0.1	-4.6	-6.7	-4.7	-5.9
6:00	10.4	6.4	9.5	7.0	8.4	6.0	3.3	0.2	-3.8	-6.6	-4.6	-5.9
7:00	10.7	6.7	9.5	7.0	8.3	5.9	3.3	0.1	-3.8	-5.9	-4.2	-5.9
8:00	10.1	5.7	9.4	6.8	7.9	4.7	3.1	-0.1	-3.3	-5.6	-3.3	-5.6
9:00	10.4	6.3	10.1	7.2	8.9	6.8	3.3	0.3	-2.6	-5.0	-2.7	-4.7
10:00	10.6	6.7	10.4	7.3	9.0	7.3	2.8	0.3	-2.0	-4.3	-2.2	-4.3
11:00	10.6	6.7	10.7	8.3	9.0	6.3	3.2	0.1	-1.6	-4.0	-1.3	-3.6
Noon	11.6	8.1	11.1	8.4	9.3	6.8	3.3	0.2	-0.2	-3.7	-0.2	-3.1
1:00 pm	12.2	8.3	11.8	9.3	9.5	7.1	3.4	0.6	0.8	-1.6	0.7	-2.7
2:00	12.3	8.4	11.4	8.9	9.5	6.7	3.5	0.9	1.4	-2.1	0.8	-1.2
3:00	11.8	7.9	11.1	7.9	9.1	6.1	3.8	0.7	1.7	-1.6	1.4	-3.1
4:00	10.8	7.3	10.8	7.9	9.0	4.0	3.9	0.7	1.7	-2.1	1.7	-2.7
5:00	11.1	7.3	11.3	9.3	8.9	5.7	3.9	1.2	1.4	-2.7	1.3	-3.1
6:00	11.3	7.3	12.3	10.7	8.9	6.3	3.9	1.4	-0.6	-4.0	-0.2	-3.1
7:00	11.3	7.2	12.7	9.8	8.9	6.3	3.8	1.0	-0.6	-4.2	-0.7	-3.7
8:00	11.7	7.5	12.0	8.9	9.5	7.0	3.8	1.4	-1.7	-5.0	-1.3	-4.0

Diurnal Temperature Measurements

COUNTY: Gong Xian County
PROVINCE: Henan
SETTLEMENT NAME: Gong Xian Town
FAMILY/HOUSE NAME: Li Songbin

House No. 11   Summer
Aug.9-10,1984
Degrees in Centigrade

Time	SITES											
	1		2		3		4		5		6	
	Dry	Wet	Dry	Wet	Dry	Wet	Dry	Wet	Dry	Wet	Dry	Wet
8:00 pm	23.8	22.9	24.2	23.9	24.5	24.0	24.0	23.8	23.9	23.7	25.6	23.9
9:00	23.5	23.3	24.0	23.9	24.4	24.2	23.9	23.9	24.0	23.9	26.4	24.8
10:00	23.4	23.1	24.0	23.9	24.4	24.1	23.9	23.7	23.9	23.4	25.1	23.7
11:00	23.4	23.0	24.0	23.9	24.2	24.0	23.8	23.7	23.9	23.9	25.6	23.9
Midnight	23.4	22.8	24.0	23.9	24.2	24.0	23.8	23.7	23.9	23.9	24.5	24.2
1:00 am	23.1	22.7	23.9	23.9	24.1	24.0	23.7	23.6	23.4	23.4	23.9	23.4
2:00	23.1	22.6	23.9	23.9	24.0	24.0	23.7	23.6	23.9	23.4	24.4	23.9
3:00	22.9	22.5	23.9	23.9	24.0	23.9	23.7	23.5	23.9	23.7	24.4	23.8
4:00	22.9	22.7	23.9	23.9	24.0	23.9	23.7	23.5	23.9	23.7	24.4	23.8
5:00	23.3	22.8	23.9	23.8	24.0	23.9	23.6	23.4	23.1	22.9	23.9	23.3
6:00	22.9	22.6	23.9	23.6	23.9	23.7	23.5	23.4	22.8	22.8	23.7	23.1
7:00	23.1	22.8	23.9	23.6	23.9	23.9	23.4	23.3	23.7	23.4	26.6	24.0
8:00	23.5	23.4	23.9	23.9	23.9	23.9	23.6	23.4	25.1	24.5	26.9	24.6
9:00	24.0	23.9	24.3	24.1	24.3	24.0	23.8	23.7	28.8	26.1	28.1	25.6
10:00	23.9	23.4	24.6	24.5	24.6	24.4	23.9	23.8	30.7	27.6	29.4	26.4
11:00	23.7	23.4	25.1	25.1	24.9	24.4	24.4	24.2	32.6	28.9	30.6	28.1
Noon	25.1	24.6	25.3	25.2	25.4	25.2	24.6	24.5	35.6	28.9	31.7	26.7
1:00 pm	25.6	25.3	25.6	25.0	25.7	25.2	24.6	24.3	37.8	29.8	32.3	27.1
2:00	24.7	24.3	25.6	25.3	26.2	25.7	25.1	25.0	35.1	29.7	32.6	28.5
3:00	24.9	24.5	25.7	25.6	26.7	26.2	25.6	25.6	33.4	28.6	32.8	28.4
4:00	25.7	25.2	26.2	25.7	26.8	26.4	26.1	25.7	31.3	28.3	33.0	28.7
5:00	25.7	25.3	25.9	25.6	26.8	26.3	26.2	25.9	29.8	27.6	33.7	28.4
6:00	24.9	24.4	25.9	25.7	26.2	26.2	26.2	25.8	28.5	27.3	33.1	28.3
7:00	24.6	23.9	25.7	25.6	26.7	26.0	26.1	25.9	27.3	26.7	30.1	28.4
8:00	24.5	24.0	25.7	25.6	26.4	25.7	25.9	25.7	26.6	26.3	28.7	27.6

Relative Humidity Calculations

COUNTY: Gong Xian County
PROVINCE: Henan
SETTLEMENT NAME: Gong Xian Town
FAMILY/HOUSE NAME: Li Songbin

House No. 11   Summer

Time	SITES					
	1	2	3	4	5	6
	RH%	RH%	RH%	RH%	RH%	RH%
8:00 pm	92.6	97.8	96.5	97.8	97.8	87.6
9:00	98.2	99.5	98.2	99.5	99.5	88.2
10:00	96.9	99.5	97.8	98.2	95.6	89.5
11:00	96.4	99.5	98.2	99.1	99.5	87.6
Midnight	95.6	99.5	99.1	99.1	99.5	97.8
1:00 am	96.8	99.5	99.1	99.1	100.0	95.6
2:00	96.4	99.5	99.1	99.1	95.6	96.1
3:00	96.8	99.5	99.5	98.6	97.8	95.6
4:00	98.2	99.5	98.6	98.6	98.2	95.2
5:00	96.4	99.1	99.1	99.1	98.2	95.6
6:00	96.8	96.9	97.8	99.1	100.0	95.6
7:00	97.7	97.3	99.5	98.6	97.8	81.4
8:00	99.1	99.5	99.1	99.1	95.7	85.2
9:00	99.1	98.2	98.2	99.1	81.4	82.2
10:00	95.6	99.5	98.2	99.1	79.2	80.2
11:00	97.8	100.0	95.7	98.6	77.2	83.3
Noon	96.5	98.7	98.3	99.5	62.2	68.9
1:00 pm	97.4	95.3	96.2	97.8	56.9	67.9
2:00	97.4	97.4	96.2	99.5	68.7	74.6
3:00	96.5	99.5	95.8	99.5	70.9	72.5
4:00	96.6	96.2	97.5	97.0	80.4	68.5
5:00	96.6	97.9	96.3	97.9	84.9	68.3
6:00	96.1	98.3	100.0	97.0	91.3	70.5
7:00	94.8	99.1	94.6	98.3	95.9	88.6
8:00	96.5	98.7	94.2	98.7	97.5	92.1

Relative Humidity Calculations

COUNTY: Gong Xian County
PROVINCE: Henan
SETTLEMENT NAME: Gong Xian Town
FAMILY/HOUSE NAME: Li Songbin

House No. 11 Winter

Time	SITES					
	1	2	3	4	5	6
	RH%	RH%	RH%	RH%	RH%	RH%
8:00 pm	61.5	71.1	43.0	59.3	67.4	58.0
9:00	59.1	64.1	64.2	61.9	67.0	67.4
10:00	60.9	64.4	74.3	67.4	68.2	67.0
11:00	60.3	64.4	70.1	60.1	73.0	70.7
Midnight	59.2	68.8	71.2	60.7	73.0	53.3
1:00 am	59.1	72.0	71.6	60.5	68.8	72.7
2:00	59.6	69.3	72.7	59.1	68.8	69.5
3:00	60.0	71.9	73.3	58.3	65.7	65.3
4:00	59.4	71.9	73.3	64.8	53.8	65.8
5:00	58.3	72.4	73.8	56.5	57.0	76.3
6:00	58.2	72.4	73.2	57.3	48.5	73.2
7:00	58.1	72.4	72.5	56.5	60.7	67.3
8:00	54.5	71.2	63.5	58.9	58.8	58.8
9:00	57.7	69.5	76.1	59.4	57.5	63.9
10:00	59.6	66.5	81.0	64.7	58.8	62.1
11:00	59.6	74.1	70.2	57.2	59.6	62.7
Noon	65.0	72.7	71.7	57.4	44.8	54.7
1:00 pm	61.5	74.4	73.1	61.2	63.5	49.6
2:00	61.7	74.6	69.5	64.2	49.4	70.8
3:00	61.6	66.7	66.8	58.8	53.8	36.7
4:00	63.6	69.6	46.2	57.6	47.0	38.9
5:00	61.3	79.0	64.2	63.4	41.1	38.4
6:00	59.0	84.0	70.7	65.5	46.3	54.7
7:00	58.4	70.9	70.7	62.5	43.0	51.8
8:00	57.9	69.2	72.4	67.6	45.8	56.5

Diurnal Temperature Measurements

COUNTY: Mang Shan Township
PROVINCE: Henan
SETTLEMENT NAME: Zhong Tou Village
FAMILY/HOUSE NAME: Liu Xueshi

House No. 12 Winter
Dec.30-31,1984
Degrees in Centigrade

Time	SITES											
	1		2		3		4		5		6	
	Dry	Wet	Dry	Wet	Dry	Wet	Dry	Wet	Dry	Wet	Dry	Wet
8:00 pm	2.9	0.1	7.0	2.0	1.2	-2.2	-1.7	-3.7	-2.0	-4.0		
9:00	2.9	0.1	9.2	2.3	1.2	-1.9	-1.6	-3.6	-1.7	-3.6		
10:00	2.9	0.3	9.1	2.4	1.3	-1.9	-1.3	-3.3	-1.1	-3.3		
11:00	2.9	0.1	6.7	3.3	1.2	-1.8	-2.2	-4.2	-2.0	-3.6		
Midnight	2.9	-0.2	6.5	3.1	1.0	-2.2	-2.2	-4.2	-2.6	-4.2		
1:00 am	2.9	-0.5	6.3	2.9	0.9	-2.5	-2.3	-4.3	-3.2	-4.6		
2:00	2.8	-0.5	6.3	3.0	0.5	-2.7	-3.6	-5.1	-4.7	-5.7		
3:00	2.8	-0.5	6.3	3.1	0.1	-3.0	-4.7	-5.7	-5.3	-5.9		
4:00	3.2	-0.8	6.3	3.1	0.0	-3.0	-4.5	-5.6	-4.7	-5.7		
5:00	2.6	-1.0	6.2	3.0	-0.1	-3.0	-4.2	-5.6	-4.3	-5.6		
6:00	2.4	-1.2	6.2	3.2	-0.3	-3.3	-4.7	-6.1	-5.0	-6.4		
7:00	2.4	-1.5	6.3	3.4	-0.5	-3.7	-5.3	-6.7	-5.6	-6.7		
8:00	1.8	-1.9	6.0	2.3	-0.8	-3.7	-4.3	-5.7	-4.7	-5.9		
9:00	1.2	-2.2	4.5	-0.9	-1.2	-3.7	-3.2	-4.7	-3.6	-5.2		
10:00	2.3	-0.7	3.9	-0.8	-0.6	-2.2	-0.7	-2.2	-0.7	-2.7		
11:00	2.7	-0.7	4.2	0.1	1.1	-1.6	0.8	-1.2	1.0	-1.7		
Noon	2.9	0.0	4.8	0.1	1.9	-1.1	1.4	-0.7	2.0	-0.2		
1:00 pm	3.1	-0.2	6.0	1.2	2.3	-0.5	2.2	2.2	2.4	0.1		
2:00	3.4	0.1	6.2	2.3	2.7	-0.2	2.8	-0.3	3.3	-0.6		
3:00	3.7	0.9	5.7	1.7	2.9	0.0	2.8	-0.3	3.3	-0.6		
4:00	3.7	0.1	6.7	2.8	2.9	0.0	2.4	-0.7	2.4	-0.7		
5:00	3.5	0.4	6.3	1.0	2.8	-0.5	1.4	-1.3	1.0	-2.0		
6:00	3.3	0.2	7.9	2.3	2.3	-0.8	1.4	-2.0	0.1	-2.0		
7:00	3.1	0.1	9.1	5.1	1.9	-1.0	-0.2	-2.2	0.1	-1.7		
8:00	3.0	0.2	9.0	5.2	1.8	-1.1	-0.7	-2.7	-0.3	-2.0		

Diurnal Temperature Measurements

COUNTY: Mang Shan Township
PROVINCE: Henan
SETTLEMENT NAME: Zhong Tou Village
FAMILY/HOUSE NAME: Liu Xueshi

House No. 12 Summer
Aug.12-13,1984
Degrees in Centigrade

Time	SITES											
	1		2		3		4		5		6	
	Dry	Wet	Dry	Wet	Dry	Wet	Dry	Wet	Dry	Wet	Dry	Wet
8:00 pm	26.0	25.0	25.6	24.5	24.9	24.6	26.5	25.2	25.6	24.9		
9:00	25.7	24.9	25.3	24.2	24.8	24.6	26.2	25.2	25.6	24.8		
10:00	25.6	24.8	25.6	24.7	24.9	24.7	26.1	25.3	25.9	25.2		
11:00	25.6	24.9	25.6	24.8	24.9	24.5	25.7	25.2	26.2	24.9		
Midnight	25.3	24.4	25.2	24.5	24.6	24.1	25.2	24.4	25.6	24.2		
1:00 am	25.3	24.4	25.1	24.2	24.6	24.2	25.2	24.4	25.6	24.2		
2:00	25.3	24.0	25.3	24.5	24.5	24.3	25.2	24.3	25.2	24.2		
3:00	25.0	23.9	25.0	23.9	24.0	23.6	24.0	23.1	23.1	22.6		
4:00	25.0	23.9	24.0	23.5	24.0	23.4	23.9	23.1	22.8	22.3		
5:00	25.0	23.8	24.7	23.7	23.9	23.3	23.8	22.9	22.8	22.3		
6:00	25.0	23.9	25.7	23.9	24.0	23.7	24.5	23.9	25.3	23.4		
7:00	24.9	23.9	24.6	23.9	23.9	23.5	24.8	23.8	25.1	23.4		
8:00	24.9	23.9	25.0	23.9	24.0	23.6	25.2	23.4	26.7	23.7		
9:00	24.9	23.9	24.8	23.9	24.0	23.7	26.2	24.0	26.7	23.9		
10:00	25.3	24.3	24.9	23.9	24.9	24.9	29.1	25.5	33.4	26.7		
11:00	25.6	25.0	25.1	24.4	25.6	25.1	35.3	27.0	30.1	25.6		
Noon	26.2	25.4	25.5	24.5	26.2	25.6	33.6	26.8	31.1	26.0		
1:00 pm	27.0	26.2	26.4	25.6	26.7	25.6	32.3	28.3	32.3	27.2		
2:00	27.4	26.2	26.4	25.6	26.8	26.2	33.5	28.2	29.1	28.1		
3:00	26.7	26.1	26.7	25.6	26.4	26.1	34.6	28.3	34.2	28.4		
4:00	26.7	26.1	26.4	25.6	26.3	25.9	31.2	27.8	28.5	27.6		
5:00	26.2	24.4	25.6	23.9	25.5	24.6	26.8	23.6	27.3	22.3		
6:00	25.9	25.1	25.6	25.1	25.3	25.0	26.3	26.2	25.1			
7:00	25.5	24.5	25.1	23.9	25.0	24.4	24.8	23.4	25.2	23.3		
8:00	25.5	25.7	25.1	23.6	24.6	24.3	25.4	25.7	25.3	21.2		

Relative Humidity Calculations

COUNTY: Mang Shan Township
PROVINCE: Henan
SETTLEMENT NAME: Zhong Tou Village
FAMILY/HOUSE NAME: Liu Xueshi

House No. 12 Summer

Time	SITES					
	1	2	3	4	5	6
	RH%	RH%	RH%	RH%	RH%	RH%
8:00 pm	92.5	91.6	97.8	90.2	94.9	
9:00	94.1	91.6	98.2	92.5	93.7	
10:00	93.7	93.3	98.2	94.1	94.5	
11:00	94.5	93.7	96.5	95.8	90.9	
Midnight	92.8	94.9	96.1	94.0	89.6	
1:00 am	92.8	93.2	96.5	93.6	93.6	
2:00	90.4	94.1	98.2	93.6	92.8	
3:00	91.5	92.0	96.5	92.6	96.0	
4:00	91.5	95.6	94.8	93.5	95.5	
5:00	91.1	91.9	94.8	93.0	95.9	
6:00	92.0	86.9	97.3	95.2	85.6	
7:00	92.4	94.8	96.5	92.8	87.5	
8:00	92.4	92.0	96.9	87.1	78.4	
9:00	92.4	93.6	97.3	83.9	80.3	
10:00	92.4	92.4	99.5	75.9	60.9	
11:00	95.3	94.9	95.8	54.1	71.3	
Noon	93.8	92.4	95.8	60.7	68.2	
1:00 pm	94.3	93.8	91.8	74.6	68.6	
2:00	91.1	94.2	95.0	68.0	92.9	
3:00	95.0	91.8	97.0	63.6	65.9	
4:00	95.4	93.8	97.5	77.7	93.2	
5:00	87.4	87.6	92.9	77.4	66.4	
6:00	93.3	95.8	97.4	99.1	91.7	
7:00	92.4	90.7	95.3	89.5	85.9	
8:00	100.0	88.7	97.4	100.0	70.4	

Relative Humidity Calculations

COUNTY: Mang Shan Township
PROVINCE: Henan
SETTLEMENT NAME: Zhong Tou Village
FAMILY/HOUSE NAME: Liu Xueshi

House No. 12   Winter

Time	SITES					
	1	2	3	4	5	6
	RH%	RH%	RH%	RH%	RH%	RH%
8:00 pm	62.2	42.2	51.7	65.7	65.3	
9:00	60.4	28.5	55.0	65.9	68.4	
10:00	62.6	30.3	53.7	66.3	62.3	
11:00	60.4	58.8	56.4	64.9	71.8	
Midnight	59.2	59.1	53.6	64.0	71.0	
1:00 am	54.9	58.2	50.2	64.6	74.1	
2:00	56.1	59.5	51.5	71.5	79.6	
3:00	56.6	61.3	52.7	79.6	87.0	
4:00	49.0	60.7	53.4	77.7	79.6	
5:00	52.4	60.6	54.1	72.7	75.7	
6:00	50.7	63.2	52.8	71.9	70.4	
7:00	47.6	64.6	49.8	70.9	75.2	
8:00	48.8	54.6	54.1	72.5	75.2	
9:00	50.9	34.4	58.5	72.0	69.4	
10:00	59.1	42.1	72.8	75.2	67.3	
11:00	55.0	46.4	60.2	70.0	59.3	
Noon	61.4	40.5	59.0	68.9	69.4	
1:00 pm	56.7	41.4	62.2	65.3	67.9	
2:00	57.4	52.3	60.9	58.2	50.7	
3:00	63.0	50.7	62.1	58.2	50.7	
4:00	52.2	53.8	61.4	57.3	57.9	
5:00	57.6	38.2	56.1	60.2	56.1	
6:00	57.4	37.9	57.8	52.3	68.4	
7:00	61.0	56.3	59.7	68.1	71.8	
8:00	60.5	58.0	59.6	67.3	73.9	

NOTES

1. Andrew Charles Hugh Boyd, *Chinese Architecture and Town Planning, 1500 B.C.–A.D. 1911* (Chicago: University of Chicago Press, 1962), 83.

2. *China Building Selection.* (Peking: China Building Information Center, 1982), 1–8.

3. • Hou Ji-Yao, "Cave Dwellings in North Shaanxi Province," *Knowledge of Architecture* 2, (n.d.), 27. (In Chinese.)

4. Qian Fu-yuan and Su Yu, "Research on the Indoor Environment of Loess Cave Dwellings" *Proceedings of the International Symposium on Earth Architecture, 1–4 November, 1985, Beijing* (Beijing: Architectural Society of China, 1985), 300.

5. Kazuya Inaba, et al., "Investigation and Improvement of Living Environment in Cave Dwellings in China." *Proceedings of the International Symposium on Earth Architecture, 1–4 November, 1985, Beijing* (Beijing: Architectural Society of China, 1985), 149.

6. Ibid., 147.

7. Jia Kunnan and Dong Yikang, "Sintered Cave for Living-Room" *Proceedings of the International Symposium on Earth Architecture, 1–4 November, 1985, Beijing* (Beijing: Architectural Society of China, 1985), 160.

8. Ibid., 161–64.

9. China Handbook Editorial Committee, *Geography*, China Handbook Series (Beijing: Foreign Languages Press, 1983), 237–39.

10. An experimental research ventilation project was conducted by the author in collaboration with the Architectural Scientific Academy of Shanxi province, Taiyuan, and Zuo Guo Bao of the academy.

11. *See* Gideon S. Golany, *Earth-Sheltered Habitat: History, Architecture and Urban Design* (New York: Van Nostrand Reinhold, 1983).

SELECT BIBLIOGRAPHY

Boyd, Andrew Charles Hugh. *Chinese Architecture and Town Planning, 1500 B.C.–A.D. 1911*. Chicago: University of Chicago Press, 1962.

Central Meteorological Department, ed. *Climate of China, Diagrams and Illustrations*. Beijing: Maps Publishing Co., 1978.

China Building Selection. Peking: China Building Information Center, 1982.

China Facts and Figures: 4,000-Year History. Beijing: Foreign Languages Press, 1982.

China Handbook Editorial Committee. *Geography*. Beijing: Foreign Languages Press, 1983.

Deng Qisheng. "Traditional Measures of Moistureproof in Raw Soil Architecture in China." *Proceedings of the International Symposium on Earth Architecture, 1–4 November, 1985, Beijing*. Beijing: Architectural Society of China, 1985.

Faegre, Torvald. *Tents: Architecture of the Nomads*. New York: Anchor Books, 1979.

Golany, Gideon S. *Chinese Earth-Sheltered Dwellings*. (Under review by press.)

————. *Earth-Sheltered Habitat: History, Architecture and Urban Design*. New York: Van Nostrand Reinhold Company, 1983. (Also published in Chinese by China Building Industry Press, Beijing, 1987.)

————. *Urban Underground Space Design in China: Vernacular Practice and Modern Lessons*. Newark: University of Delaware Press, 1989.

History of Chinese Architecture. Group compilation. Beijing: China Building Industry Press, 1979. (In Chinese.)

Hou Ji-Yao. "Cave Dwellings in North Shaanxi Province." *Knowledge of Architecture* 2 (n.d.). (In Chinese.)

Inaba, Kazuya, et al. "Investigation and Improvement of Living Environment in Cave Dwellings in China." *Proceedings of the International Symposium on Earth Architecture, 1–4 November, 1985, Beijing*. Beijing: Architectural Society of China, 1985.

Jia Kunnan and Dong Yikang. "Sintered Cave for Living-Room." *Proceedings of the International Symposium on Earth Architecture, 1–4 November, 1985, Beijing*. Beijing: Architectural Society of China, 1985.

Liu Duanzheng. *Chinese Housing Concepts*. Beijing: Architectural Construction Press, 1957. (In Chinese.)

Qian Fu-yuan and Su Yu. "Research on the Indoor Environment of Loess Cave Dwellings." *Proceedings of the International Symposium on Earth Architecture, 1–4 November, 1985, Beijing*. Beijing: Architectural Society of China, 1985.

Rapoport, Amos. *House Form and Culture*. Englewood Cliffs, N.J.: Prentice Hall, 1969.

FIGURE CREDITS

1. Redrawn after the following sources: (A) Torvald Faegre, *Tents: Architecture of the Nomads*. New York: Anchor Books, 1979, 82; (B, C, F, and H) *History of Chinese Architecture*, group compilation. Beijing: China Building Industry Press, 1979, 128, 126, 122, and 117 respectively; (D and G) Liu Duanzheng, *Chinese Housing Concepts*. Beijing: Architectural Construction Press, 1957, 131 and 132 respectively; (E) Deng Qisheng, "Traditional Measures of Moistureproof in Raw Soil Architecture in China," in *Proceedings of the International Symposium on Earth Architecture, 1–4 November, 1985, Beijing*, Beijing: Architectural Society of China, 1985, 66. (All in the Chinese language except A and E.)

4. Central Meterological Department, ed. *Climate of China, Diagrams and Illustrations*. Beijing, China: Maps Publishing Co., 1978.

5. Ibid.

11. After G. S. Golany, *Earth-Sheltered Habitat*. New York: Van Nostrand Reinhold, 1983.

12. After Amos Rapoport. *House Form and Culture*. Englewood Cliffs, NJ: Prentice Hall, 1969.

Table 2. Architectural Scientific Academy of Shanxi, with the collaboration of Zuo Guo Bao, architect.

INDEX

161